NEW NATURALIST LIBRARY

Limpback Reprints

MOUNTAIN FLOWERS

When originally published in 1956, John Raven and Max Walters' *Mountain Flowers* was the first book on British mountain flora and it has remained a definitive title ever since. One of the earliest botanical works to lead readers familiar with the European Alps to appreciate British alpine flowers, it is still a first-class companion today. Increasing urbanisation and the spread of motorways etc have changed Britain's mountain areas less than other habitats, despite the numerous visitors who seek recreation in them, so that, except for minor changes of taxonomy, a problem familiar to all botanists, the text requires no revision to be as valid now as when it was written.

Written by a professional botanist, Dr Walters, and an amateur, the late John Raven, an authority on Greek philosophy, it is a combination of the best in academic and field research. It is based upon the author's joint experience of flower-hunting in the hills of Scotland, Ireland, Wales and the Lake District – as well as drawing upon their wide knowledge of the scientific and historical background of Britain's mountain flora. The text and line illustrations are reproduced unaltered from the edition of 1965 but the process of manufacture used to achieve an economic price does not, unfortunately, do full justice to all the photographs; and those originally in colour appear in black and white. Dr Walters has contributed a new introduction.

First published 1956

Limpback edition 1984
ISBN 0 00 219070 2

Printed and bound by Biddles Ltd, of Guildford and King's Lynn

COLLINS
GRAFTON STREET LONDON

Preface to the Limpback Edition

It gives me great pleasure to contribute some introductory words to this reprint of *Mountain Flowers,* not least because the exercise enables me to recall many very happy hours of botanising with John Raven, my co-author, who died in 1980. In particular I remember details of our joint tour of Scotland in the late Summer of 1953, planned in part to provide material for the book, which appeared first in August 1956.

As many reviewers have rightly pointed out, the text of *Mountain Flowers* shows very clearly indeed the disparity of style between the professional and the amateur, a disparity of which John and I were uncomfortably aware. The fact that the two approaches could co-exist, and even come to influence each other, might, however, be claimed to illustrate precisely what the 'New Naturalist' series aims to achieve: namely, an integration of the scientific and the aesthetic approach to natural history. I would not wish to iron out these differences, even if the exigencies of photographic reproduction permitted any change.

What of the changes which thirty years of botanical study of the British mountain flora have brought about? Do they seriously date the book? My considered opinion is that, provided the reader is informed of one or two important reference works published since 1956, both the general background picture and the particular appeal of the book are still unimpaired.

Undoubtedly the most important 'new' book is the second edition of Professor Sir Harry Godwin's monumental work, *History of the British Flora* (Cambridge University Press, 1975). (In Chapter 4, we refer the reader for more information to the first edition of Godwin's book, published, like *Mountain Flowers,* in 1956). In the 1975 edition much new information is summarised for individual species of our mountain flora in the form of distribution maps incorporating both modern and subfossil records: a particularly well-documented example is *Potentilla crantzii,* which is clearly a relict British mountain plant more widespread in Late Glacial times.

In addition, an excellent general account of the history of our flora is now available at a price most readers could afford: Pennington, W., *The History of British Vegetation* (English Universities Press, 1969) and Alastair Fitter's *Altas of the Wild Flowers of Britain and Northern Europe* (Collins, 1972) provides a very useful source of maps of British mountain plants showing their adjacent Scandinavian range.

The standard work on *The Pollination of Flowers* by Michael Proctor and Peter Yeo, a splendid addition to our New Naturalist series published in 1973, discusses in a wider context the subject raised in our review of the biology of the mountain flora, namely the tendency shown by mountain plants towards self-pollination and apomixis.

So much for new books. What should we say about detailed new knowledge of our mountain flora? Surprisingly, perhaps, there have been no major discoveries of 'new' (i.e. overlooked or undiscovered) British mountain plants since the spate of remarkable new records in the early 1950s – *Artemisia norvegica, Diapensis lapponica, Koenigia islandica* – all of which were reported in our book, though the Irish mountain flora has the new and unexpected discovery of the alpine species *Minuartia recurva*. On the other hand, Don's Twitch *(Agropyrum donianum),* whose re-discovery was recorded so entertainingly by John Raven in our book, is now no longer considered to be worthy of specific distinction from the common British grass *Agropyrum caninum,* and has suffered the further ignominy of being transferred to the genus *Elymus!*

S.M.W., 1984

THE NEW NATURALIST

A SURVEY OF BRITISH NATURAL HISTORY

MOUNTAIN FLOWERS

The aim of this series is to interest the general reader in the wild life of Britain by recapturing the inquiring spirit of the old naturalists. The Editors believe that the natural pride of the British public in the native fauna and flora, to which must be added concern for their conservation, is best fostered by maintaining a high standard of accuracy combined with clarity of exposition in presenting the results of modern scientific research.

THE NEW NATURALIST

MOUNTAIN FLOWERS

by

JOHN RAVEN

M.A.

FELLOW OF KING'S COLLEGE, CAMBRIDGE, AND
UNIVERSITY LECTURER IN CLASSICS

and

MAX WALTERS

M.A., PH.D.

CURATOR OF THE UNIVERSITY HERBARIUM,
CAMBRIDGE

WITH 44 PHOTOGRAPHS
AND 20 DISTRIBUTION MAPS

COLLINS
14 ST. JAMES'S PLACE LONDON

First Published 1956
Reprinted 1965

TO

CHARLES RAVEN

BUT BECAUSE Mountains have been lookt upon by some as Warts and superfluous Excrescencies, of no use or benefit; nay, rather as signs and proofs, that the present Earth is nothing else but a heap of Rubbish and Ruins, I shall deduce and demonstrate in Particulars, the great Use, Benefit, and Necessity of them . . .

For it is a true Observation, that Mountains do especially abound with different Species of Vegetables, because of the great diversity of Soyls that are found there, every Vertex, or Eminency, almost affording new kinds. Now these Plants serve partly for the Food and Substenance of such Animals as are proper to the Mountains, partly for Medicinal Uses; the chief Physick-Herbs and Roots, and the best in their kinds growing there: it being remarkable, that the greatest and most luxuriant Species in most Genera of Plants are Native of the Mountains: partly also for the Exercise and Diversion of such ingenious and industrious Persons, as are delighted in searching out these Natural Rarities; and observing the outward Form, Growth, Natures, and Uses, of each Species, and reflecting upon the Creator of them his due Praises and Benedictions.

RAY, J. 1692. *Wisdom of God*
Ed. 2, pp. 199 and 203.

CONTENTS

PLATES IN COLOUR*

It should be noted that throughout this book Plate numbers in Arabic figures refer to Colour Plates, while Roman numerals are used for Black-and-White Plates.

*As explained on the first page, these plates are here reproduced in black and white.

PLATES IN BLACK AND WHITE

EDITORS' PREFACE

Very little has been written about our mountain flowers, in comparison with those of other British habitats. This is a curious fact. Professor Pearsall, in his *New Naturalist* volume *Mountains and Moorlands*, dealt fully, of course, with the ecology of mountain vegetation; but a detailed treatment of the individual species was outside the scope of his book. The main reason for this neglect is probably the great popularity among British holiday-makers of the European Alps. Their unrivalled botanical attractions have tended to overshadow the less spectacular home-grown displays in the British Isles. But anyone who knows and loves the flora of our mountains will agree that this neglect is unjustified, and will (we feel sure) welcome a volume that, for the first time, deals fully with British mountain flowers.

The joint authorship of Mr. John Raven and Dr. Max Walters epitomises, in a sense, the present very healthy state of British field botany. Dr. Walters is a professional botanist; Mr. Raven is one of our leading amateurs; and their collaboration in planning and writing the present volume exemplifies the close and beneficial integration that now exists between professional and amateur botanical investigation in this country. Max Walters is Curator of the University Herbarium at Cambridge and is director of the five-year scheme, recently launched by the Botanical Society of the British Isles (with the support of the Nuffield Foundation and the Nature Conservancy) for producing maps showing the distribution of British flowering plants. He has already collaborated with the Botanical Editor in the *New Naturalist* volume *Wild Flowers*, and is one of the leading workers on the experimental study of British plants. John Raven, son of Canon C. E. Raven, is a Fellow of King's College, Cambridge, and an authority on Greek philosophy. During the last few years he has made a special study of our mountain flora, especially in Scotland, and has concentrated on that fascinating group of plants, the hawkweeds (*Hieracium*), which attains its greatest and most puzzling diversity in mountain habitats. As will be seen from the pages that follow, the two authors have carried out several joint expeditions to British mountains and have shared many of the experiences that they describe so vividly. John Raven, like his father, is a skilled botanical artist and, between them, they have now painted nearly the whole British flora, including many species of hawkweed never illustrated in colour before.

Our general aim in the *New Naturalist* series is to present each subject from as broad a point of view as possible, and we feel that the authors of the present volume have fulfilled this aim to a notable degree. In Part I they sketch the scientific background to the British mountain flora, and show how its treasures have gradually been discovered by botanists, famous and obscure, during the last four hundred years. The stories of the unexpected additions to the flora that have recently been made are particularly intriguing. In Part II detailed descriptions of the flora of selected mountains are given, based on the first-hand knowledge of the authors. In the Appendices, the keys to the mountain species of *Alchemilla, Hieracium* and *Salix* should stimulate more botanists to take an interest in these difficult groups, so often shunned by amateur (and professional!) botanists.

This latest volume in the *New Naturalist* series will, we hope, lead many who are familiar with the European Alps to realise more fully the botanical attractions of our own mountains, and, perhaps, to contribute towards a solution of some of the scientific problems that their flora presents.

THE EDITORS

AUTHORS' PREFACE

In these days, when the mountains of Britain are so readily accessible, it is not surprising that an interest in mountain flowers is widely shared by amateur and professional alike. This book is designed, not as a technical, scholarly and strictly scientific work—for Professor Pearsall's earlier volume in this series has admirably provided such a study of the vegetation of mountains and upland moors—but rather as a guide and a stimulus to those many naturalists who think of the mountain flora primarily in terms of its individual species; those who long to see, or perhaps treasure the memory of seeing, the purple saxifrage on the high cliffs of Ingleborough or the blue sow-thistle on the inaccessible ledges of a Highland glen.

We are aware that no strict definition of a "mountain plant" is possible, and we shall not attempt to justify all our inclusions or omissions. We have, however, tried to mention, if only in the lists in the Appendix, all species of vascular plants which often occur on British mountains above 2,000 feet; and we have paid special attention to those whose total world distribution includes either the arctic regions, or the main mountain ranges of Europe, or both. Part II in particular, in which we describe and discuss the characteristic species of certain selected districts, makes no claim to be an exhaustive account of the British mountains. Some areas, such as the Brecon Beacons in Wales and the mountains near Dumfries or those around Loch Torridon in Scotland, receive hardly a mention: neither of us is sufficiently familiar with them to be prepared to describe them in any detail. Others, such as Snowdonia, Upper Teesdale and even the whole of Ireland, have already been fully described in earlier books that are readily available: in such cases we refer the reader to the books in question and do no more than touch upon particularly relevant topics. The selected areas which we have discussed in more detail are, we believe, fairly representative of the British mountains as a whole. Our accounts of them are arranged, so far as possible, in an order from the most acid, and therefore floristically poorest, to the most alkaline, and therefore richest. We have, however, deliberately altered this order in one respect. It seemed best to deal with Snowdonia and the English Lake District before starting on the Scottish Highlands, even though the Cairngorms afford the most extreme examples of vegetation characteristic of acid rock.

Many plants which grow on British mountains, and especially rushes, sedges and grasses, have no English names. In such cases there is clearly nothing to be done but to call the plant in question by its Latin name. But in general our policy on this thorny question has been to treat each chapter as a separate unit, at the first mention of a particular species in any chapter to give both its English and its Latin names, and thereafter to refer to the plant in question by the one or the other but not by both. All the Latin names which we use, and very nearly all the English (but here we have used a few more which seem to us tolerable), follow the nomenclature adopted in the recent *Flora of the British Isles,* by A. R. Clapham, T. G. Tutin and E. F. Warburg (1952. Cambridge).

In the case of some of the rarest or most local plants we have been purposely vague in our statements about locality and habitat. This is a regrettable necessity; for although it is probably true to say that there is today less collecting of rare "alpines" by covetous botanists or gardeners than there was in the last century (when the mountain ferns in particular were sadly diminished), nevertheless the danger unfortunately persists, and irreparable damage can be done by the few. It is our earnest hope that our own and future generations can learn, before it is in some cases too late, to treasure our rich heritage of mountain flowers, and, whilst rightly enjoying the pleasures of the chase, sublimate the instinct for the kill.

One of us being a professional botanist and the other an amateur, our approach to the subject is naturally different. Each of us has written his own chapters in his own words. The reader would probably have little difficulty in determining for himself where the one of us takes over from the other; but in order that he should be in no doubt at all where the blame attaches, we have headed each chapter or section with the initials of its author. In the case of Chapter 1, however, the initials are misleading. Some years before we ourselves began to write this book, it was hoped that Dr. H. R. Fletcher (of the Royal Botanic Garden at Edinburgh) would be its author; and Dr. Fletcher did get so far, before he was compelled to abandon the project, as to collect a vast amount of material for such a book. As soon as he learnt that we were to take his place, Dr. Fletcher handed over to us, with a generosity for which it is difficult to render adequate thanks, all the results of his researches, and, in particular, his material for the historical survey in Chapter 1.

We should like also to record our grateful thanks to the Editors and Publishers of the series; to the many expert photographers who have put their stock of photographs at our disposal; to Mr. R. W. David and Dr. C. D. Pigott, who very kindly read the proofs for us and have both helped us with valuable suggestions; and last, but by no means least, to the many friends, too numerous to name, who have accompanied us on the various expeditions to the mountains which this book records. As Professor J. H. Balfour wrote, in the passage quoted *in extenso* in the Introduction, "the companionship of those who are prosecuting with zeal and enthusiasm the same path of science is not the least delightful feature of such excursions."

J.R.
M.W.

INTRODUCTION (J.R.)

ONE SUNNY DAY early in September, 1951, four botanists set out from Killin determined to see, in the only district where it was then known to grow, the very rare mountain grass, *Agropyron donianum*. The walk was short, easy and entirely successful. That evening, back in Killin, one of the four, the expert on grasses at whose instigation the whole expedition had been planned, remarked thoughtfully but with obvious relief: "I shall never have to climb a mountain again." So far as I know, he has remained true to his resolve: certainly the next day, while the other three spent many arduous hours on another and ultimately successful search for the same grass high on Ben Lawers, he pottered gently up the banks of the River Lochay. I know what he meant, too, about botanising on mountains; in the peculiar despondency that marks the end of a long and unsuccessful search in the hills I have sometimes even reached the same resolve as he did. But the next day I have gone up the mountain again with as much hope and enthusiasm as ever.

Some botanists, I suppose, are so catholic in their tastes that an expedition, say, to the Lizard Head or to the Breckland will afford them neither more nor less pleasure than an expedition to Perthshire or Angus. Others may share with the expert on grasses his marked preference for the level lowlands. Indeed there are even some whose favourite hunting-ground seems to be the municipal rubbish-dumps of the larger cities, with their annual crop of alien weeds. But there can be little doubt that a Gallup Poll of all the botanists in Britain would result in an overwhelming majority in favour of the mountains. The reasons are obvious enough, and have been appreciated by generations of British botanists.

A little more than a century ago, in the summer of 1847, Professor J. H. Balfour, of Edinburgh, led a party of his students on a three-weeks' excursion to the two areas of Scotland most famous for their mountain flora, the Clova district and Ben Lawers. His account of this excursion, which appeared in the *Edinburgh New Philosophical Journal* for July, 1848, contains the following eloquent preface: "Excursions may be truly said to be the *life* of the botanist. They enable him to study the science practically, by the examination of plants in their living state, and in their native localities; they impress upon his mind the structural and physiological lessons he has received; they exhibit to him the geographical range of species, both as regards latitude and altitude; and with the pursuit of scientific knowledge, they combine that healthful and spirit-stirring recreation which tends materially to aid mental efforts. The companionship too of those who are prosecuting with zeal and enthusiasm the same path of science, is not the least delightful feature of such excursions. The various phases of character exhibited, the pleasing incidents that diversified the walk, the jokes that passed, and even the very mishaps or annoyances that occurred—all become objects of interest, and unite the members of the party by ties of no ordinary kind. And the feelings thus excited are by no means of an evanescent or fleeting nature; they last during life, and are always recalled by the sight of the specimens which were collected. These apparently insignificant remnants of vegetation recall many a tale of adventure, and are associated with the delightful recollection of many a friend. It is not indeed a matter of surprise that those who have lived and walked for weeks together in a Highland ramble, who have met in sunshine and in tempest, who have climbed together the misty summits, and have slept in the miserable sheiling—should have such scenes indelibly impressed on their memory. There is, moreover, something peculiarly attractive in the collecting of alpine plants. Their comparative rarity, the localities in which they grow, and frequently their beautiful hues, conspire in shedding around them a halo of interest far exceeding that connected with lowland productions. The alpine *Veronica* displaying its lovely blue corolla on the verge of dissolving snows; the Forget-me-not of the mountain summit, whose tints far excel those of its namesake of the brooks; the *Woodsia* with its tufted fronds adorning the clefts of the rocks; the snowy Gentian concealing its eye of blue in the ledges of the steep crags; the alpine *Astragalus* (Pl. II, p. 19) enlivening the turf with its purple clusters; the *Lychnis* [i.e. *Viscaria*] choosing the stony and dry knoll for the evolution of its pink petals; the *Sonchus* raising its stately stalk and azure heads in spots which try the enthusiasm of the adventurous collector; the pale-flowered *Oxytropis* confining itself to a single British cliff; the Azalea forming a carpet of the richest crimson; the Saxifrages with their white, yellow, and pink blossoms clothing the sides of the streams; the *Saussurea* and *Erigeron* crowning the rocks with their purple and pink capitula; the pendent Cinquefoil blending its yellow flowers with the white of the alpine Cerastiums and the bright blue of the stony *Veronica*; the stemless *Silene* giving a pink and velvety covering to the decomposing granite; the yellow *Hieracia* whose varied transition forms have furnished such a fertile cause of dispute among botanists; the slender and delicate grasses, the chickweeds, the *Carices*, and the rushes, which spring up on the moist alpine summits; the graceful ferns, the tiny mosses, with their urn-like thecae, the crustaceous dry lichens with their spore-bearing apothecia, all these add such a charm to Highland botany as to throw a comparative shade over the vegetation of the plains."

It is strange that one who knew and loved his mountain plants so well should have so cheerfully turned them into "apparently insignificant remnants of vegetation." No doubt his herbarium was accessible to his students and served an educational purpose. In these days certainly there would be less emphasis on collecting, and especially on collecting such plants as the alpine milk-vetch (*Astragalus alpinus*) or the red mountain catchfly (*Viscaria alpina*), the known localities of which are so few that a single specimen in the main public herbaria should suffice. Yet even now, more than a century later, the collector remains a menace to our flora. If the herbarium, and especially the private herbarium, had never been invented, there would be appreciably more specimens of the *Lychnis* evolving their pink petals on the stony and dry knoll, the clefts of more rocks would be still adorned with the tufted fronds of the *Woodsia,* and that secretiveness which rightly characterises so many of the best field-botanists would be a needless foible.

But with that one important proviso, Professor Balfour surely speaks the truth. Doubtless the modern Professor of Botany, even when he has led his students on such an excursion, would seldom allow himself so glowing a digression or his readers so revealing a glimpse of his own personality. But if Balfour's style is dated his sentiments are not. Few British botanists will rest content until they have seen the sights he describes; and few, when once they have seen them, will not agree that they "throw a comparative shade over the vegetation of the plains." For a century and more the great mountain masses of

Britain, and especially of Scotland, have exercised on botanists an unrivalled fascination.

And yet, even in so small an area as the British Isles, there is still much to be done. Vast tracts of country, notably in the Western Highlands of Scotland, still await a thorough botanical exploration; and such an exploration can and should be undertaken by the amateur as well as the professional. True, it is necessary first to acquire a fair working knowledge of the British mountain flora. It is not only essential to know the plants by sight, it is also highly desirable to have developed a flair for the right kind of locality; and for that purpose beginners are still probably best advised to go first to the most famous hills. But once that stage is over—and it need not take very long—then the real fun begins. For great as may be the pleasure of seeing on Ben Lawers the many botanical treasures for which it has long been famed, it still does not to my mind compare with that of finding even such relatively common flowers as mountain avens (*Dryas octopetala*) or northern rock-cress (*Cardaminopsis petraea*) in a district from which they have never before been recorded. It is largely in the hope of stimulating further exploration of the remoter mountains of Britain that this book has been written.

Part One

GENERAL

CHAPTER I

THE PIONEERS (J.R.)

ON JULY 22ND, 1639, Thomas Johnson, Apothecary and Royalist, editor and emendator of Gerard's *Herbal,* left London with two companions, Paul Sone and Edward Morgan, for North Wales. First they journeyed to Chester, where they were met by Walter Stonehouse, a diligent searcher after rare plants in Yorkshire. Then, by way of Flint, Conway, Pen-maen-mawr and Bangor, the party reached Caernarvon twelve days after leaving London. On August 3rd they ascended Widhfa, called by the English Snowdon.

"Here, among the other high mountains, the peak raised its head as does the Cypress among the Viburnums. Desirous of seeing the peak we hurried on and obtained the services of a country boy as guide, because not only the summit but the whole mountain was wrapped in cloud. Then, leaving our horses and outer garments, we began to climb the mountain. The ascent is at first difficult, but after a time the way broadens out—though still equally steep, with vast precipices on the left and a difficult climb on the right. After negotiating three miles, we at last reached the crest of the mountain, shrouded in thick cloud. Here the path is very narrow and, as we climbed it, we were appalled at the rugged crags and precipices on both sides, and also the Stygian marshes here and there, the largest of which is called by the natives the Devil's Home. At the summit we sat down in the clouds. First we arranged in order the plants collected at our peril among the rocks and precipices, then ate the food we had brought with us."

This passage leaves some doubt whether Johnson and his friends did actually climb to the summit of Snowdon. It was long ago pointed out that the *stygiae etiam hinc et inde paludes, quarum maxima Daemonis domicilium ab incolis vocatur,* might well refer to the Llyn Idwal area, in which case *Daemonis domicilium* itself would obviously be the "Devil's

Kitchen", the name still given to the great chasm on Glyder Fawr, the mountain opposite Snowdon. But be that as it may, the fact remains that in the written record of this horrifying ascent we have for the first time a list of the plants of one of our British mountains. It is not a bad list either: among the twenty plants that were considered noteworthy are northern rock-cress (*Cardaminopsis petraea*), mountain sorrel (*Oxyria digyna*), the roseroot (*Sedum rosea*), the moss campion (*Silene acaulis*), the alpine saxifrage (*Saxifraga nivalis*), the starry saxifrage (*S. stellaris*), the alpine saw-wort (*Saussurea alpina*) and least willow (*Salix herbacea*), as well as (to Johnson's surprise) such maritime plants as the thrift (*Armeria maritima*) and the sea campion (*Silene maritima*). A modern botanist, even if perhaps fearing less for his neck, might well reap a less rich harvest from a day on the Welsh hills.

Even before Johnson's time there are, of course, published references to our mountain plants. Thomas Penny, for instance, who was born about 1530, had collected in the North of England such plants as the bird's-eye primrose (*Primula farinosa*) and the cloudberry (*Rubus chamaemorus* Pl. I, p. 18)—the latter, according to De l'Ecluse (who was sent it by Penny and published an admirable account of it in his *Stirpium Pannonicarum Historia*, (pp. 117-19), growing "in great plenty among heather on Mount Ingleborrow, the highest in all England, twelve miles from Lancaster." Similarly, James Cargill, a doctor of Aberdeen, who had lived in Switzerland with Jean Bauhin and been a pupil of Bauhin's brother Caspar at Basle, had sent to Bauhin in 1603 the first recorded British specimen of the chickweed wintergreen (*Trientalis europaea*), which Bauhin duly reported in his *Prodromus* of 1620 (p. 100) as "Pyrola Alsines flore Major. . . In betuletis Scotiae natam D. Cargillus ex Scotia misit." But such records, even when they are of plants of the mountains, are few and isolated; however keenly Dr. James Cargill may have botanised in the birch-woods of Scotland, there is no surviving evidence that he ever botanised on a Scottish mountain.

William How's *Phytologia Britannica*, which was published in 1650 and was for nearly two and a half centuries regarded as the first attempt at a complete enumeration of our flora, is now known to be little but a revised version of Johnson's *Mercurius*. The records from the *Mercurius* are, however, augmented by others from several of How's friends. Walter Stonehouse, fellow-traveller with Johnson on the expedition to Snowdon, reappears in the *Phytologia*; he had provided

How with a record of *Primula farinosa* "most plentifully mixt with Pinguicula on a very low and squalid meadow near Knasborough"; and Richard Heaton, one of the first students of Irish botany, is responsible for records of the spring gentian (*Gentiana verna*)—"Gentianella Alpina verna. . . . In the mountaines betwixt Gort and Galloway abundantly"—and of the mountain avens (*Dryas octopetala*)—"Teucrium Alpinum Cisti flore. . . . In the mountaines betwixt Gort and Galloway : it makes a pretty show in the winter with its rough heads like Viorna"—that is, traveller's joy (*Clematis vitalba*).

When How died in 1656, his publishers invited Christopher Merret to compile a complete catalogue of the flora, fauna and fossils of Britain. Merret was to have been assisted in his task by a skilled botanist named John Dale; but on the death of Dale in 1662 he secured instead the services of an old soldier, Thomas Willisel. Merret's *Pinax*, which was published in 1666, contains in its list of plants not a single reference to Willisel; but a comparison of the new records in the *Pinax* with the plants ascribed to Willisel in Ray's *English Catalogue* of 1670 makes it appear virtually certain that many of the north-country records in the *Pinax* are in fact Willisel's. There is, for instance, alpine penny-cress (*Thlaspi alpestre*), "in the pastures above the ebbing and flowing well two miles from Giflewick in stony ground amongst the grass" (p.118)—a record that is shown to be most probably Willisel's by the fact that Ray also saw the plant, in 1668, "above the ebbing and flowing well a mile from Settle"; there is the hoary whitlow-grass (*Draba incana*), "at Clapdale in Yorkshire in the midway betwixt Settle and Ingleborough hill on the rocks"; and there are similar records of several other plants which Ray was shown in the same localities by Willisel, including such mountain plants as purple saxifrage (*Saxifraga oppositifolia*), "on the North side of Ingleborough hill, near a bog by the side of an underground river", or alpine bistort (*Polygonum viviparum*), "at Crosby in Westmorland and Ravenswaith; and two miles from Settle, Yorkshire, in the pastures towards Ingleborough hill in a place called Cromock." The district around Settle and Ingleborough, which was a favourite hunting-ground of the early botanists, seems in fact to have been first thoroughly worked by Thomas Willisel.

The great John Ray himself, of course, made many an excursion into the British mountains. In 1658 he visited first Derbyshire and then North Wales. In Derbyshire he saw alpine scurvy-grass (*Cochlearia *alpina*) growing in some abundance near Castleton; and in Wales, besides "divers rare plants" on Snowdon, "de quibus consule catalogum," he found on Cader Idris the globe-flower (*Trollius europaeus*), the alpine clubmoss (*Lycopodium alpinum*) and the lesser clubmoss (*Selaginella selaginoides*). In 1660 he toured the North of England, accompanied by his friend and pupil Francis Willughby. The following year, with the same companion, he ventured into Scotland, where "they have neither good bread, cheese or drink," "their butter is very indifferent, and one would wonder how they could contrive to make it so bad," "they have rarely any bellows or warming-pans" and "the people seem to be very lazy, at least the men, and may be frequently observed to plough in their cloaks." In 1662, yet again with Willughby, he revisited North Wales, paid a hasty visit to Carnedd Llewelyn, "which signifies Llewelyn's bones, a very high hill", and found on Snowdon, among other things, the parsley fern (*Cryptogramma crispa*), "which we had before observed in Westmorland", four club-mosses and the mossy saxifrage (*Saxifraga hypnoides*). Again in 1668, after three summers on the Continent and two in the South of England, he spent a fortnight botanising, this time alone, in Yorkshire and Westmorland. It was on this tour that he found "near Orton in Westmorland, by a stream running across the road to Crosby", the hitherto undescribed alpine bartsia (*B. alpina*). Finally in the early summer of 1671, although the first edition of his *Catalogus Angliae* was already published, he returned yet again to the same districts, the flora of which had obviously impressed him. This time, with the aid of Thomas Willisel, he saw alpine bistort (*Polygonum viviparum*), alpine lady's mantle (*Alchemilla alpina*), vernal sandwort (*Minuartia verna*), dwarf cornel (*Chamaepericlymenum suecicum*), shrubby cinquefoil (*Potentilla fruticosa*) and many other interesting plants; and when he so far overcame his distaste for Scotland as to venture a little beyond the border, he was rewarded for his fortitude by the first discovery, "about two miles from Berwick, by the side of a rivulet, in a boggy ground", of the Scottish asphodel (*Tofieldia pusilla*). All these plants he recorded in the second edition of the Catalogue, which appeared in 1677.

Ray's thirty-two page supplement to the Catalogue, the *Fasciculus Stirpium Britannicarum post editum Plantarum Angliae Catalogum observatarum,* was published eleven years later. His own excursions had by this time been severely curtailed, chiefly by responsibility to his family, and the majority of the records contained in the supplement had been sent him by other botanists. At least one of these contributors, Edward Lhwyd, was himself an unusually keen and active naturalist; among other records attributed to him are alpine mouse-ear chickweed (*Cerastium alpinum*) from Snowdon, rock cinquefoil (*Potentilla rupestris*) from Craig Breidden in Montgomeryshire, and alpine meadow-rue (*Thalictrum alpinum*) from Cader Idris. Our knowledge of his early botanical activities begins with a letter, written in 1682, in which he gives directions to David Lloyd for an exploration of the Welsh hills. On Aran Benllyn, he says, can be found two species of saxifrage (*S. stellaris* and *S. hypnoides*), cowberry (*Vaccinium vitis-idaea*) and fir club-moss (*Lycopodium selago*); on Craig Verwin *Lycopodium alpinum*; on Snowdon, in addition to a number of plants already recorded from there, the dwarf juniper (*Juniperus communis* ssp. *nana*); and on Cader Idris not only rose-root, mountain sorrel and lesser clubmoss, but also brittle bladder-fern (*Cystopteris fragilis*) and crowberry (*Empetrum nigrum*).

Even more interesting than this list are the detailed instructions in the same letter for collecting on Cader Idris. Though the vasculum had yet to be invented, the menace of the collector was already rife.

"To employ some trusty fellow yt shall undertake to traverse Cader Idris for a whole day in search of plants, observing punctually these following directions.

Takeing a handbasket with him, he must go up as far as Llyn y Cau ere he takes notice of anything: but then he must trace some rivulet of water, as high as he can with safety, putting into his basket ten or a dozen roots of each sort of herb he can discover; but of such as are very small, 15 or 20. He must sometimes stray from the rivulets to the rocks and gather anything he meets with there; wch he found not near the rivulets. He must not omit any sort of plant yt he sees; excepting ye common sort of ffern, Heath and grasse.

Of all ye shrubs he meets with: I mean such as Llûs duon, Llûs Cochion, Gruglys, Mwyar y Mynudh, Crâch-helig &c.; let him take up onely 3 or 4 roots and those the least. Let him not neglect any sort of Mosse: such as Troed y blaudh, Corn y Carw, Mwswgl y ffynod-wydh, Mwswgl y Cypreswydh (wch creeps abundantly on ye grasse towards the Top of the Hill) &c. He must be cautious in picking up the very least thing his eyes can discover; for by ye rills of water there are some plants soe small as scarce to be seen, which neverthelesse are as rare as any.

Wee don't confine him to the rivulets all day; for when he has search'd several of them, he may wander amongst the rocks, pastures &c. on all sides of the Hill; but let him gather nothing that grows lower than a quarter of a mile of ye Top; & let him be sure that he'l finde the greatest variety by ye rivulets of water and other wet place; espec. in Craig dhu above Llyn y Cau; where he must not fayl to climb as high as he can with safety. ffor his further encouragement, when we shall receive these plants, we shall send him by ye carrier a farthing for every different kind he has gather'd; wch if he proves any thing diligent, can not amount to less than 2 shillings.

If there be noe possibility of sending to Cader Idris let these directions be observed at Rennogl Fawr".

Lhwyd did not always do his botanising by proxy; there can be no doubt that he knew his Welsh mountains very well. By the summer of 1689 he was sending his collections to Ray, whose main contribution to British botany, the *Synopsis Stirpium Britannicarum*, was to appear the following year (1690), with a second edition, Ray's last major work on the British flora, in 1696. Lhwyd's most notable discovery was unquestionably the plant which still bears his name, *Lloydia serotina*. In the 1696 edition of the *Synopsis* it was first described as follows: "Bulbosa alpina juncifolia pericarpio unico erecto in summo cauliculo dodrantali: I have not seen the flower and cannot tell to what group of bulbous plants it belongs: on the stalk are two or three leaves rather narrow and short: on the highest rocks of Snowdon." But Lhwyd did find flowers in plenty in 1696, and the specimens he sent to Ray evoked the comment: "The bulb with a single flower which you had seen in seed before, if it be not a plant *sui generis* but reducible to any of the known kinds, I think it may be referred to Ornithogalum".

Nor were Lhwyd's botanical activities confined to Wales; in 1700 he is writing to Ray's great friend, Dr. Tancred Robinson, of the plants he has found on the mountains of Ireland. "In the same neighbourhood," he writes, "on the mountains of Ben Bulben and Ben Buishgen, we met with a number of the rare mountain plants of England and Wales, and three or four not yet discover'd in Britain"; and one of these three or four, which was collected near Sligo in 1699 and preserved in the herbarium of the Rev. Adam Buddle, was none other than the endemic Irish sandwort (*Arenaria ciliata* ssp. *hibernica*). Again, "On the mountains of Keri, *Sanicula guttata* [i.e. *Saxifraga spathularis*, St. Patrick's cabbage] grows in abundance; together with some other rare plants", including starry saxifrage, alpine saw-wort and alpine lady's mantle. And finally "*Pentaphylloides fruticosa* [i.e. *Potentilla fruticosa*] we found plentifully amongst lime-stone rocks on the banks of Loch Crib in the County of Galloway; and Dr. Merret's *Vaccinia rubra foliis Myrtinis crispis* [i.e. the bearberry, *Arctostaphylos uva-ursi*] (a very beautiful plant) we found to be no rarity in this kingdom." The list is admittedly short, but it is a landmark in the history of Irish botany. Apart from Heaton's records in How's *Phytologia,* hardly anything else was yet known of the mountain flora of Ireland: and for the next century, despite some records from the Mourne Mountains in Dillenius' edition of Ray's *Synopsis* (1724), others from the Wicklow Mountains in the appendix to Caleb Threlkeld's *Synopsis Stirpium Hibernicarum* (1726), and lists of plants in Charles Smith's Histories of Kerry, Cork and Waterford, no contribution to the subject compared in value with that of Lhwyd.

Ray died in 1705, Lhwyd in 1709. Between them, and with the assistance of others such as Willisel, they had amassed a knowledge that could almost be called intimate of the mountain flora of both England and Wales. Ireland too had yielded up some of its treasures. Scotland alone, or at any rate the Highlands, still awaited exploration. True, Robert Sibbald, aptly termed the Father of Scottish Botany, had already in 1684 published his *Scotia illustrata*; but although it mentions some five hundred plants, almost the only notable mountain species is that which is still called after him, *Sibbaldia procumbens.* Neither did the next ninety years witness anything more than isolated records, such as Dr. Walker's discovery, in 1761, of the purple oxytropis (*O. halleri*). Hudson's *Flora Anglica,* first published in 1762 and an excellent compilation of existing knowledge of the British flora, contains only a few Scottish records.

Our knowledge of the flora of the Scottish mountains may be said to date from 1778, when, despite the fact that its title page is dated 1777, the Rev. John Lightfoot's *Flora Scotica* was first published. Lightfoot's journey through Scotland had been undertaken in 1772, at the invitation of the celebrated zoologist and antiquary Thomas Pennant, who travelled with him. In the preface to the Flora, Lightfoot tells how the two companions journeyed "either by land or sea from the South of Annandale to the borders of Sutherland, visited most of the Hebrides, except the Long-Island, traversed the Kingdom from Argyleshire to Mearns, that is, from the western to the eastern shore, and afterwards return'd to England by way of Edinburgh and Kelso; that in all this tract, which took up the daily exercise of a whole summer, I had a constant eye to the following work, embraced every opportunity of scaling the highest mountains, climbing the most rugged rocks, penetrating the thickest woods, treading the fallacious bogs, winding upon the shores of seas and lakes, in short, of examining every variety of land or water, which promised to produce a variety of vegetables." The detailed itinerary is not given, but much of it can be easily pieced together. On the way north to Sutherland they visited the Isle of Arran, where the few-flowered sedge (*Carex pauciflora*) was added to the British flora, and the islands of Bute, Islay, Jura, Colonsay, Mull, Iona, Rhum and Skye. On the mountains of Rhum, especially Barkeval, they saw a number of interesting plants, including the mossy cyphel (*Cherleria sedoides*); Skye yielded, among other things, an alpine hawkweed, the frog orchis (*Coeloglossum viride*) and the moonwort (*Botrychium lunaria*). From the islands they came to Loch Maree; then to Dundonnel and the hills at the head of Little Loch Broom, where they found dwarf cornel (*Chamaepericlymenum suecicum*) and the bog whortleberry (*Vaccinium uliginosum*); and so at last, by way of the Knockan rocks, where they gathered *Dryas*, the holly fern (*Polystichum lonchitis*) and the green spleenwort (*Asplenium viride*), to Tongue in Sutherland. The return journey too was equally productive, especially their exploration of the Perthshire mountains. Thanks largely to the guidance of the Rev. John Stuart of Killin, who was not at that time ordained to a charge, they discovered, and so for the first time made known, the astonishingly rich flora of the Breadalbane range. Ben Lawers itself is strangely absent from the Flora: it would appear that even Stuart had never explored it. But Creag na Caillich, Meall Ghaordie, Meall nan Tarmachan and Beinn Heasgarnich are among the familiar names which appear in Lightfoot's pages.

Obviously the *Flora Scotica* marks a huge advance in the knowledge of the flora of Scotland. It lists in all—that is, including phanerogams and cryptogams—some twelve hundred and fifty species, of which well over fifty are flowering plants of the mountains. Yet even so it met with severe criticism. As Pennant says of it, no sooner did the Flora appear than "envy emptied her whole quiver, she employed every little archer to aim her envenomed dart and, unhappily, with too much success. The book received its condemnation". Pennant himself withdrew all available copies of the book and "imprisoned them for a few years in a darksome cell of a London warehouse". There they remained until 1792, when, envy's whole quiver being evidently not inexhaustible, Pennant ventured to reissue them under the somewhat misleading title of "second edition". It is a depressing commentary on human nature that a topic so apparently peaceable as the flora of Scotland should give rise, as from time to time it has, to such envenomed controversy.

Lightfoot himself, in the summer after his Scottish tour, turned his attention to the mountains of Wales, where he seems to have kept fairly closely to the beaten track. But others were soon to follow him onto the Scottish hills. James Dickson for instance, a Covent Garden nurseryman, paid two visits to Scotland, in 1789 and 1792, and seems to have been the first to reveal the riches of Ben Lawers. His loot on the first occasion included the rock speedwell (*Veronica fruticans*), the rock whitlow-grass (*Draba rupestris*) and the boreal fleabane (*Erigeron borealis*); the second visit added the drooping saxifrage (*S. cernua*) and the snow gentian (*G. nivalis*). No doubt the gentian was already, as we know it was some fifty years later, prudently "concealing its eye of blue in the ledges of the steep crags." Not only the owners of private herbaria have improverished Ben Lawers; the owners of private rock-gardens are a greater threat still. They can unfortunately claim for their predatory practices the spurious respectability of a long lineage.

Meanwhile, from 1779 onwards, another and more famous nurseryman had been eagerly exploring the Highlands, the great George Don himself. His *Herbarium Britannicum*, dedicated to Sir Joseph Banks, then President of the Royal Society, began to appear in 1804; from then onwards four fasciculi, each of twenty-five plants and containing a due proportion of rare mountain species, were to be issued yearly. His claims in the preface are probably just. "Since he first began his botanical excursions into the Highlands of Scotland, in the year 1779, he is confident (and he hopes he may mention it without the imputation of vanity) that he has traversed more of the Caledonian Alps than any other botanist has ever done. He has repeatedly ranged over the great mountains of Angusshire which surround the great district of Clova, where no one on a similar pursuit has ever preceded him. He has also searched the vast range of mountains which stretch about sixty miles through the district of Knoydart, in Inverness-shire, a region which had never before, nor has since, been examined with a botanical eye. He is the only botanist, too, who has explored the lofty mountains of Cairngorm and the great hills of the neighbourhood." Nor indeed, as a list of his discoveries will show, are the districts he mentions by any means his only fields of activity.

On Ben Lawers itself Don added several more to Dickson's records. The rare sandwort, *Minuartia rubella,* the beautiful alpine forget-me-not (*Myosotis alpestris*) and that most attractive of all British sedges, *Carex atrofusca,* were new to Britain. Hardly less notable were the fern, *Woodsia alpina,* and the rushes, *Juncus biglumis* and *J. castaneus.* On Meall Ghaordie he discovered *Bartsia alpina,* the first record for Scotland, and the alpine pearlwort (*Sagina saginoides*). *Carex vaginata* was his best find on Cairngorm, the curved woodrush (*Luzula arcuata*) on the great bulk of Ben Macdhui. Time and time again he explored Clova and was the first to make known to botanists this rich and lovely district. Here he found *Carex rariflora* and the alpine hawkweed, *Hieracium calenduliflorum,* both of which were new to science; three more of his discoveries—another hawkweed (*H. lingulatum*), the yellow oxytropis (*O. campestris*) and the red alpine catchfly (*Viscaria alpina*)—were new to the flora of Britain. Ben Nevis yielded, among other things, the very rare grass, *Poa flexuosa;* on Ben Lomond he saw alpine mouse-ear chickweed and *Carex saxatilis,* on Ben Voirlich purple saxifrage and on Schiehallion *Carex vesicaria* and alpine meadow-rue. His many ascents of Lochnagar revealed for the first time the unusual interest of its flora; his records from there include not only one species new to science, the alpine foxtail (*Alopecurus alpinus*), but also such rarities as brook saxifrage (*S. rivularis*), blue sow-thistle (*Cicerbita alpina*) and the very rare hybrid grass *Poa x jemtlandica.* Even if we ignore for the moment those of Don's records which have never been confirmed, we can still accept the verdict of G. C. Druce (1912): "his discoveries of new plants are probably larger and more important than those made by any other British botanist."

Don, unlike Dickson, was a Scot; his nursery garden was in Forfar. In *The Scots Magazine* for June 1809, there is an interesting account of this garden by Don's friend, Dr. Patrick Neill. "The existence of a flower garden and flower nurseries at Forfar, which for numbers, diversity and rarity of the hardy plants cultivated in it are perhaps scarcely to be surpassed in Britain, is a fact not generally known. We think it right to give it what publicity is in our power, both as a piece of interesting information to botanical amateurs, and of justice to the indefatigable exertions of Mr. George Don, who, we understand, has surmounted many difficulties in following out his favourite pursuit, and in forming so extensive and curious a collection of living plants. The whole of the plants are of a hardy sort, Mr. Don not possessing either green-house or stove for the protection of such as are tender. It is in *alpine plants* and in *hardy perennials,* and *annuals,* that the Forfar garden excels. The garden is situated on a bank which slopes down to the lake of Forfar, not far from the town; and it fortunately includes a great variety of soils, from dry to peat bog. No place could be found more favourable for alpines and aquatics, which are in general found to be of rather difficult cultivation, but which flourish here as in their native habitats. . . . To give some idea of the extent of the collection, I shall mention the number of species of several genera which are at present growing in the garden. Of the genus *Veronica,* there are 55 species, of *Salvia* 50 species, *Campanula* 44, *Allium* 40, *Saxifraga* 46—some of the rarest ones, as *S. caesia, S. petraea, S. rivularis,* etc.; *Dianthus* about 20 species, *Cucubalus* 13—being the whole ever cultivated in Britain; *Silene* nearly 50, *Fumaria* 14, the genera *Ononis, Lathyrus, Vicia* almost complete, *Astragalus* 40 species, *Trifolium,* no fewer than 69, *Hieracium* 44. It were needless to enumerate more. The botanist will form a due estimate of this collection on being told that he may see here upwards of 60 species of *Carex,* flourishing in great perfection. The agriculturist may here find the whole of the hardy Gramina, carefully distinguished and arranged, amounting to over 100 kinds. This season Mr. Don has introduced several hundred species of hardy plants, most of which we are told have never before been cultivated in Scotland. Among the rare British plants at present in flower in this garden may be mentioned the elegant little grass called *Knappia agrostidea* [i.e. *Mibora minima*] . . . and the *Holosteum umbellatum.* . . .

"The Forfar garden, it must, however, in conclusion, be confessed, makes very little external show, being in a great measure destitute of the ornament which arises from neat alleys with hedges or edgings, or well laid-out or well-kept gravel walks. It is, in fact, merely an uncommonly excellent collection of hardy plants; and while it would doubtless fail to please the lover of tasteful gardening, it would as certainly prove highly interesting to the botanist and to the curious cultivator. Mr. Don, we have been told, has an ample nursery of rare hardy plants, for which he receives orders from the curious in different parts of Britain; and when the proceeds of these shall enable him, we understand it to be his intention to improve the exterior appearance of his garden."

This long description of the Forfar garden is not as irrelevant as it may at first sight appear. There is no denying that among Don's many first records there are some which not only have never yet been confirmed, but seem on the whole unlikely ever to be confirmed in the future. Such, to cite two well-known instances, are *Ranunculus alpestris* "about 2 or 3 rocks on the mountains of Clova" or *Potentilla tridentata* from the hill of Werron. In the middle of last century, of course, many records were too readily dismissed as erroneous which have since been fully substantiated. Arnott especially, Professor of Botany at Glasgow, carried incredulity to extremes; he wrote, for example, of *Viscaria alpina* on Little Culrannoch: "We have strong reasons for thinking that the plant was sown there about sixty years ago." Since this and other such sentences were written, Don has many times been vindicated; nor is it safe to assume, in view of the recent rediscovery of the alpine coltsfoot (*Homogyne alpina*, see p. 182), that he will not be vindicated yet again. But when all is said, it still does seem likely that Don made a number of inadvertent mistakes. No doubt his garden was in no wise different from other botanic gardens. If his labels were occasionally misplaced, or one species invaded the territory of another, then he may well have recorded as British a number of species which had been sent him from elsewhere. He still remains the greatest of all the Scottish pioneers. In the words of an inscription in the churchyard in Forfar, "so long as there are students of the Alpine flora of Britain, his name will be held in affectionate remembrance."

Don died in 1814, a little more than a century after Lhwyd. Lightfoot and he had done for the mountain flora of Scotland what Ray and Lhwyd did for England and Wales. Henceforth for all three countries there was a substantial framework of knowledge into which new discoveries could be duly fitted. Ireland alone still awaited thorough exploration. True, in 1794 Walter Wade, a Dublin doctor, had published what may justly be considered the earliest Irish County Flora, the *Catalogus systematicus plantarum indigenarum in comitatu Dublinensi inventarum,* and had followed it up ten years later with a comparatively full list of the rarer plants he had seen in Ireland, chiefly in Connemara, Cork and Kerry. But that these represented a mere beginning he was himself aware. Discussing, in the preface to the latter

J. Markham

Plate I. Cloudberry, *Rubus chamaemorus*

M. C. F. Proctor

Plate II. Alpine milk-vetch, *Astragalus alpinus*. Glen Clova publication, the need for a complete Flora of Ireland, he observes: "I am aware that there are some genuine and valuable materials for such a Flora, in the hands of a gentleman in the northern part of this kingdom, a person every way qualified, from industry, information and acuteness, for such an undertaking, and it is much to be lamented he does not gratify the botanical world with the result of his researches. Not having the honour of his acquaintance, or pleasure of his confidence, I shall not presume to mention his name; but I shall hope, when his leisure permits, that he will follow my example, by publishing the habitats of such valuable and scarce plants as he has met with growing wild in Ireland—from such a source, the communication must be highly interesting."

The gentleman in question, whose name Wade does not presume to mention, was John Templeton of Belfast. Whilst George Don was assiduously hunting the Scottish countryside for rare plants to grow in his Forfar garden, Templeton was also busily employed in cultivating a garden which he had himself laid out in 1793. "This had been partly an orchard, partly an osier ground; and conducting through it a stream of water, raised on an artificial rock, he rendered it in every way fit for the intended purpose. Here he collected, from various parts of the world, rare and useful plants, which he endeavoured to naturalise in this climate, by placing them in a soil and situation resembling, as nearly as possible, that to which they had been accustomed. By this means there is growing in his garden, in the open air, a wonderful and curious collection of plants from India, China, North and South America, Siberia etc.; such as *Camellia japonica, Thea viridis* (the tea plant), *Ailanthus praecox,* and others, which were formerly kept in the hot house, and then in the greenhouse. Most of the trees at Cranmere (all, I believe, except the chestnuts and oaks) were raised from seed or planted by himself; and so great a variety of the natives of the forest has, perhaps, never been collected in so small a place" (Drummond, 1829).

Like his Scottish counterpart, however, Templeton was by no means exclusively concerned with cultivated plants. He was also, as Wade suggested, extremely well informed of the plants of his own country; he was a frequent contributor to the *English Botany* of Smith and Sowerby; and he was the original finder of many of the rarer plants of the North East of Ireland, including such mountain species as *Dryas, Selaginella selaginoides* and *Silene acaulis.* In 1793, he began to compile a catalogue of Irish plants, an octavo volume bearing the title, in his own hand, "Catalogue of the Native Plants of Ireland observed by John Templeton, A.L.S."; and from 1794 until 1814 he was continually adding to it new localities, almost entirely from the North East. For many years this manuscript copy was the only work of his known to exist. But he also kept a Journal, of which, though the parts before 1806 are lost, thirteen octavo volumes survive, covering the period from 1806 onwards to a mere seven months short of his death in 1825. In the journal are numerous notes of the excursions he made and of the plants he found. Moreover, Wade was not misinformed when he suggested that Templeton had the materials for a projected *Flora Hibernica.* There were indeed six quarto volumes in cardboard covers, in which, among other things, were beautiful watercolour drawings of the Mosses, Liverworts and Seaweeds.

Templeton's project eventually came to nothing: both he and Wade died in 1825. That very year, however, saw the publication of the *Catalogue of the Plants found in Ireland*, by James Townsend Mackay, the first attempt at a complete list ever published; and though in 1833 Katherine Bailey, later Lady Kane, had published *The Irish Flora,* which was largely based on the records of John White, of the Glasnevin Botanic Garden, it was Mackay who, in 1836, first produced the comprehensive Flora that Wade and Templeton had contemplated. His *Flora Hibernica* sums up all the knowledge that had yet been accumulated of the flora of Ireland.

For the best part of two hundred years, then, from the time of Johnson to the second quarter of the nineteenth century, much of the initiative of British field-botanists was directed towards the exploration of the mountains. The plants were there on the hills awaiting discovery and Lhwyd, Ray, Lightfoot, Don, Templeton and many others, could not fail to make important new additions to the number of plants already known. By 1830, when Sir William Hooker published the first edition of his *British Flora,* the number of British flowering plants was already estimated to be over fifteen hundred. There were others, of course, still waiting to be found, especially perhaps on the mountains, and some, such as the alpine milk-vetch or the very rare sedge, *Carex norvegica,* did not for much longer escape detection. But the pioneering stage was over, the days of specialisation were beginning, and any history of British botanists must henceforth become either arbitrarily selective or intolerably tangled. In the present context there are perhaps two lines of development which especially deserve mention ; but it must be remembered that there are many other botanists besides those mentioned who have contributed, some indeed who still are contributing, to our knowledge of the flora of the British mountains.

The first notable development in the nineteenth century is the rapidly increasing interest in the problems of geographical distribution and in ecology. Hooker himself indeed, in the preface to his Flora, significantly asserts that he is "rather anxious to indicate the range of the species than the precise spot where any one is to be found." But Hooker was not the first in this field. Nathaniel Winch had already, in 1819, at a meeting of the Literary and Philosophical Society in Newcastle-on-Tyne, read a very remarkable paper entitled "*An Essay on the Geographical Distribution of Plants through the Counties of Northumberland, Cumberland and Durham*", in which for the first time a particular portion of the British flora is analysed and classified according to geographical distribution. And what Winch did for plant geography William MacGillivray soon did for ecology. His "*Remarks on the Phanerogamic Vegetation of the River Dee,*" in which he traces the mountain vegetation by gradual stages downwards from the summits to the valleys, is the earliest of all British ecological studies.

Both Winch and MacGillivray expressed the hope that their fragmentary studies would stimulate others to prepare a complete picture of the geographical distribution of British plants. In this hope they were not to be disappointed. Very soon after the publication of MacGillivray's short paper, indeed within the same year of 1832, a small volume was printed for private circulation with the title "*Outlines of the Geographical Distribution of British Plants,*" by Hewett Cottrell Watson. In 1845 came a celebrated paper by Edward Forbes discussing the origin of the mountain flora and fauna (see p. 70). And in 1847 appeared the first volume of Watson's *Cybele Britannica,* which, along with his *Topographical Botany,* is not only the answer to the prayers of Winch and MacGillivray, but also the basis of every subsequent study of the distribution of British plants. Watson's influence will be evident enough in later chapters of this book. For the moment it must suffice to quote one short passage only from the *Cybele,* the relevance of which is obvious :—

"The question may hence be asked by readers why the term *arctic* is now employed for the regions and zones, instead of the more familiar term *alpine*; this latter term having been adopted in the author's earlier writings on the distribution of plants. Two considerations induce to the change of term. First, the word *alpine* is in common use to express mountain plants, without reference to their positions above or within the limits of cultivation; and it has thus a different and less precise signification than the word *arctic*, as here applied and restricted. Secondly the flora of all our mountains above the limit of cultivation has much closer affinity with that seen northwards of the British islands than it has with the Alps of Middle Europe; it is, for example, much more like the flora of Lapland than the flora of Switzerland."

The second notable characteristic of the nineteenth century is the increasingly intensive study of the difficult or critical genera. A comprehensive account of the detailed taxonomic work done at this stage in the history of British botany would entail prolonged research and in the end fill a whole volume. The scurvy-grasses, the pearlworts, the willows, the sedges and the grasses, genera well represented on British mountains, were all of them, among several other groups, beginning during the century to engage the critical attention of the best British botanists. By the end of the century not only was the main task accomplished of identifying the British species of these groups with the Scandinavian or Alpine, but a considerable number of new species had been based, by British botanists, on British material. Not merely from personal preference, but because the hawkweeds are, of all the critical genera, much the most widespread on the hills of Britain, I shall take as the most appropriate example of this detailed taxonomic study the genus *Hieracium*.

George Don himself, of course, even before the beginning of the nineteenth century, had been considerably interested in British hawkweeds. Not only was he the first to discover certain mountain species, but he had also cultivated, as Dr. Neill explicitly mentions in his description of the Forfar garden, as many as forty-four species of this one genus. It was, however, the two Backhouses who first drew wide attention to the taxonomic problems involved. From 1845 onwards, father and son together made repeated visits to Scotland, chiefly in search of hawkweeds; and in 1856 the son, James Backhouse Junr., duly published his *Monograph of the British Hieracia*. This monograph actually contains, in contrast to the two hundred and sixty of its latest successor, a mere thirty-three species, of which eleven were then new.

But it remains the first book published in this country that is devoted to the one genus only. James Backhouse Junr. can in fact be regarded as father, in Britain at least, of the "splitters" as opposed to the "lumpers". Unfortunately (or, as some may think, wisely), as soon as the Monograph was written, Backhouse himself seems to have lost interest in the genus. Although he continued for over thirty years to make authoritative pronouncements on the specimens referred to him, his utterances bore a steadily diminishing relation to reality. But until his death in 1890 there seems to have been nobody who was prepared, on this subject at least, to gainsay him.

The great days of the hieraciarchs, as H. W. Pugsley calls them, were around the turn of the century. F. J. Hanbury, the Rev. E. S. Marshall, the Linton brothers (both parsons), W. H. Beeby, the Rev. Augustin Ley and the Rev. W. H. Purchas were all at this stage busily engaged, in the field, the garden and the study, on the genus *Hieracium*; with how much effect, can be easily judged from the spate of papers on the subject, unparalleled before or since, which appeared in the *Journal of Botany* in the years between 1889 and 1911. Hanbury's *Illustrated Monograph of the British Hieracia,* the first part of which was published in 1889, came unfortunately to a premature end; the speed at which new species were being added to the list seems to have discouraged him at a relatively early stage. But the Rev. W. R. Linton's *Account of the British Hieracia,* which appeared in 1905, was only effectually superseded when, in 1948, H. W. Pugsley's *Prodromus of the British Hieracia* was posthumously published in the *Journal of the Linnaean Society of London.* And if there is still any question of the thoroughness with which these seven enthusiasts, and others like them, had tackled their task, then it is only necessary to see, in Pugsley's *Prodromus,* how many species were first found and recorded by one or other of the seven.

Nor was any of the seven merely a hieraciarch; each was a general field-botanist of the highest order. Indeed it is not in connection with a hawkweed at all, but in describing a new mountain species of scurvy-grass, *Cochlearia micacea,* that the Rev. E. S. Marshall published his eloquent defence of the "splitters." "The extreme difficulty of this variable genus," he wrote in the *Journal of Botany* (1894), "is well known to students, and I should not have ventured to put forward the present plant as distinct, but for its essential constancy when grown for some years under totally changed conditions, and its coming true from seed. Forms of *alpina* of Watson, under the same circumstances, remain about equally constant, as do *danica, officinalis* and *groenlandica.* Anyone is, of course, free to take into account only an aggregate "*C. polymorpha*" corresponding to the Benthamic "*Rubus fruticosus*", but this treatment is hardly calculated to increase knowledge or promote accuracy. If such utterances as Mr. N. E. Brown's crude and off-hand dismissal of the *Epilobium* hybrids in the Supplement to *English Botany* Ed. 3, or Sir J. D. Hooker's laconic condemnation of the Hanburian *Hieracia,* in the popular edition of *Island Life,* are to be taken as representing the views and teaching of our botanical Gamaliels, it becomes a serious question whether or no "critical" plants are worth studying at all. For my own part, I think that field-botanists have some ground for complaint, when careful and deliberate conclusions, arrived at as the result of prolonged research in the open, the garden, and the study, are hastily tossed aside as worthless, without their properly investigating the matter, by those whose expressions of opinion deservedly carry great weight, and whose reasoned criticisms would be very valuable. This ultra-conservative tendency, while it has doubtless spared us some errors, has hitherto kept our knowledge of British plants from reaching the standard of completeness attained on the Continent." That is the pronouncement not of a mere hieraciarch but of a thorough botanist. It may be that the particular plank on which Marshall chose to balance his argument is not the strongest he could have found; the specific distinctness of *Cochlearia micacea* is still an open question. It may be again that the passion for splitting is even now carried to needless extremes; certainly the present number of British hawkweeds or brambles is calculated to deter all but the rabid enthusiasts. But even if in the end the pendulum swings back the other way, still there will be no denying that the movement stemming from Backhouse has greatly increased our knowledge and understanding of the British flora.

Meantime of course, even while these two particular tendencies were at their height, there was still an incalculable number of botanists who went to the mountains simply to see what they could see; and some of them, such as G. C. Druce or R. Lloyd Praeger, added as much to our knowledge of the distribution of British plants as any of their illustrious predecessors. In Ireland especially there was intense activity; H. C. Hart, R. W. Scully, N. Colgan, S. A. Stewart and many others were all at work simultaneously on the thorough exploration of the Irish mountains and rivers, the lakes, islands and coasts. Remarkable men they were too, combining with their expert botanical knowledge an irresistible sense of adventure and fun. The following episode, for instance, which was recounted by Hart's lifelong friend R. M. Barrington in the *Irish Naturalist* (Barrington, 1908), seems to have been characteristic of both the men involved. "By appointment he [Hart] turned up at Fassaroe one dreadful day to botanise on the cliffs around Powerscourt Waterfall, and to hear, if possible, the Wood-Wren. Well knowing that if the expedition failed the incident would for years afterwards be a theme for ridicule, a few slices of bread were hastily wrapped up, and we started in torrents of rain, absolute silence being observed regarding the atmospheric conditions. Both wet to the skin 'in no time', Hart deliberately kept walking among the scrub, briars and long grass by the river's edge, so as to discourage his companion. To prove utter indifference to moisture, the writer walked into the river and sat down on a submerged stone, and began to eat lunch. Hart, with the utmost nonchalance and without saying a word, did likewise. Saturation was soon complete. All rivalry ceased, and friendship prevailed during the remainder of the day." The flora of the British mountains can occasionally, as a number of incidents have already shown, be a surprisingly fertile source of controversy and recrimination. For such occasional acrimony these Irishmen's memoirs are the best antidote.

Despite MacGillivray's pioneer study of British mountain vegetation, which was published as early as 1832, interest in the mountain flora still remained, throughout the greater part of the nineteenth century, definitely floristic rather than ecological in its bias. It is interesting to find, however, that several of the earliest vegetational studies in the modern ecological tradition were made by workers on the northern and mountain floras. Particularly relevant in the present context are the studies of R. Smith (1900) on North Perthshire, of his brother W. G. Smith (1904) on Forfar and Fife, and of Crampton (1911) on the vegetation of Caithness. The work of the Smiths, indeed, to quote Professor A. G. Tansley in the preface to the famous *Types of British Vegetation* (1911), "formed the nucleus of the material dealt with in (this) book." Even the account of arctic-alpine vegetation in Tansley's standard work on British vegetation (1938) is still largely based on W. G. Smith's account; for almost the only considerable ecological study of mountain vegetation published between those years was that of Price Evans (1932) on Cader Idris. It seems as if the study of mountain vegetation, which got off to a flying start in the Smiths' *Botanical Survey of Scotland* and similar work, has somehow failed to fulfil its early promise: the British literature still lacks detailed vegetational studies of large areas of mountainous country. There is reason to hope, however, that as the work of the Nature Conservancy progresses, many of these gaps will be gradually filled.

Even in the second half of the twentieth century, to cut a long story as short as possible, the botanical exploration of the British hills is not at an end. In 1950 came the news of *Koenigia islandica* on the hills of Skye; though it had been discovered in 1934, it had remained undetected in the material of *Peplis* at Kew (see p. 135). The summer of 1951 witnessed first the refinding of "one of George Don's reputed discoveries", the alpine coltsfoot (*Homogyne alpina*) (see p. 182), and then a most notable addition to the British list in the shape of *Diapensia lapponica* (see p. 137). And finally, in 1952, what is possibly an even more notable addition was made to the records of the Scottish mountain flora, the very rare species, related to the wormwoods, *Artemisia norvegica* (see p. 138). Ten years ago a British botanist might well have felt and claimed that the list of our mountain plants was now complete. Recent events have demolished that claim; and if there is anything to be learnt from the way in which these recent discoveries have been made, it would seem that in the remoter regions of the Scottish mountains the initiative of the naturalist, amateur as well as professional, may even yet be richly rewarded.

CHAPTER 2

PHYSICAL FEATURES AND VEGETATION OF BRITISH MOUNTAINS (M.W.)

THERE IS no rigid definition of a 'mountain', but we can for convenience restrict our use of the term to land over 2000 ft. As the map (fig. 1, p. 28) shows, all the high ground in the British Isles is situated in the north and west, where the older and harder rocks are to be found. All these main mountain areas are geologically speaking of a great age, far older than the relatively recently-formed Alps, for example. This is the explanation of the familiar rounded contours of most British mountains, which are in such striking contrast to the spectacular giants of the Alps; the forces of nature which combine gradually to lower and to smooth the surface have been at work for so long on the ancient British mountains that many of the sheer cliffs and steep edges have been smoothed down, whilst in the Alps the erosion of the main peaks is still by comparison furiously active.

On British mountains the steep, unstable slopes and cliffs are mostly recently-formed and often well below the rounded summits; they are the products of the glacial period, when virtually the whole of Britain, with the exception of the south of England, was at some time covered by immense ice sheets. The familiar glaciated valleys, U-shaped in section, of the northern countryside owe their contours to the deep scouring by the glaciers they once contained; and with the disappearance of this ice, the lower slopes of the mountains have been left in an unstable, 'over-steepened' condition with lateral 'hanging valleys' terminating above the main ones and often emptying impressive waterfalls down the steep slopes. At the heads of such glaciated valleys may be fine corries (Welsh 'cwms') with more or less vertical rock faces and typically a small stony lake in the bottom (Pl. III, p. 30). On such steep slopes erosion by water, frost and ice may be very active.

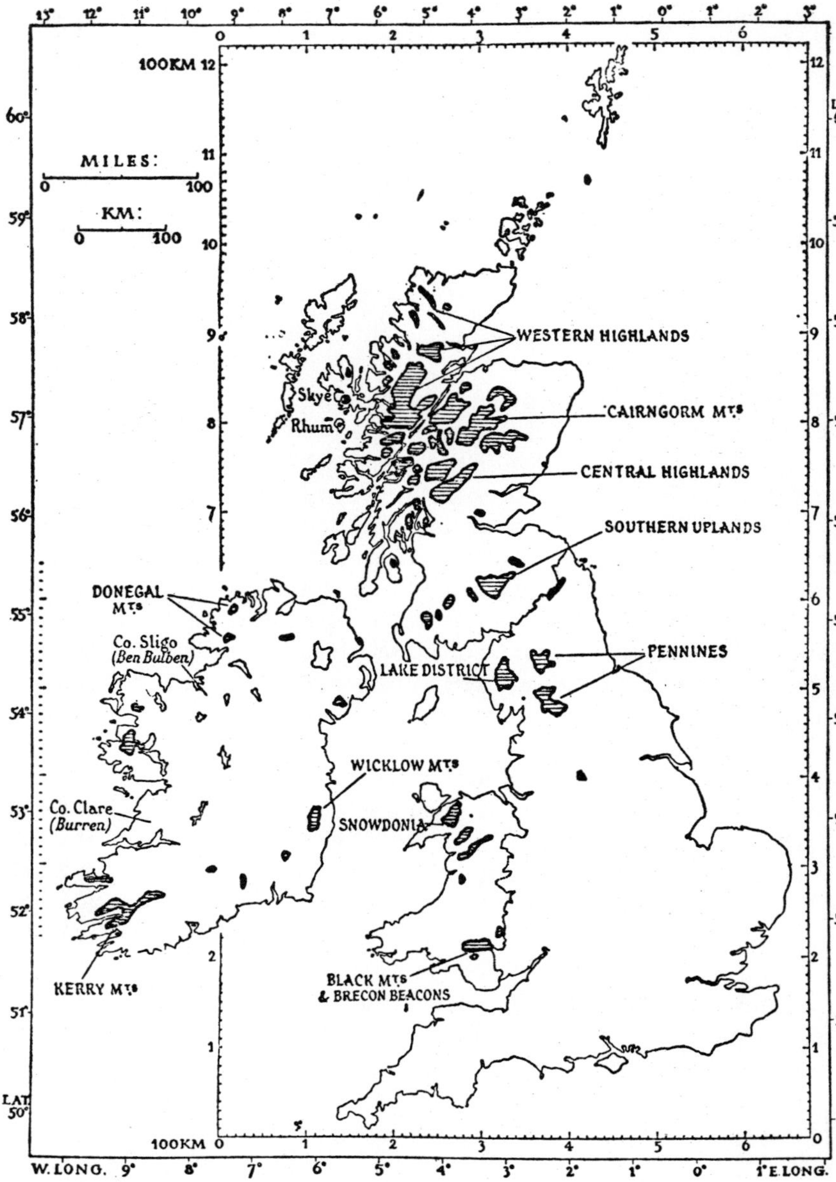

FIG. 1
Map of British Isles showing distribution of main areas of land over 2,000 ft.

Thus, in spite of the great age and consequent maturity of British mountain masses, there is, largely owing to the geologically recent glaciations, a considerable amount of steep unstable rock, scree and immature soils. Further, over many flatter upland areas, the ice-sheet has scraped away the previous soil, and left bare rock over which a vegetation cover may only partly be established; and enormous quantities of glacial debris (boulder clay) have been plastered over the lower ground in the mountain masses, as well as widely over the surface of lowland Britain. These results of the ice age have a most important bearing upon the distribution of the mountain flora, for different elements have been able to survive under the diverse conditions so provided.

In order to gain some understanding of the mountains as habitats for a remarkable flora, we must look at the physical and chemical properties of the parent rock and the overlying deposits. It must be a familiar fact to most field-botanists that the rich variety of mountain plants is to be found in relatively few areas, often very restricted, on British mountains. Some of these areas—for example the Ben Lawers range in Central Perthshire—have long been deservedly famous for their alpine flora, and thousands of botanists have made their pilgrimages there. Another fact, initially rather surprising, about the mountain flora of Britain is that it is by no means necessarily at great altitudes. Indeed, the summits of the 4000-foot mountains in Scotland harbour remarkably few different kinds of plant. What seems to determine the presence of a rich flora more than anything else is the chemical content of the rock or soil, and particularly the presence in quantity of bases such as lime, potash and magnesia. Those mountain areas such as the Lawers range which are composed of soft basic mica-schists provide habitats for a great variety of plants which cannot grow on the acid granites, gneisses and schists which make up so much of the Scottish Highlands. Hardness of rock is of importance, too; thus the Black Cuillins of Skye, composed of a *hard* basic igneous rock (gabbro), have not as a whole as rich a flora as the lower cliffs of the softer basaltic mountain range in the north of the island (see Ch. 7). The effect of local pockets of basic rock in an acidic mountain region may be very spectacular; examples of this are given in later chapters.

It is difficult to generalise about the climatic conditions on British mountains; but one thing at least can be said. Precipitation, in the form of rain or, in winter, snow, is generally high, and this is of the greatest importance, when coupled with the relatively low summer temperatures and high atmospheric humidity, in its effects upon mountain soils. These conditions favour the development of water-logged, boggy soils with a strongly acid reaction; on flat, badly drained ground thick 'blanket bog' peat will easily develop, whilst in any better-drained condition, the high rainfall will cause a rapid 'leaching' or dissolving out of the soluble bases, which, as we have seen, seem to be essential for the growth of many mountain plants. Although high rainfall is generally characteristic of British mountains, there is, of course, very considerable local variation, as between, for example, the Cairngorms in the Eastern Highlands and the much wetter Western Highland ranges; or more locally between eastern and western Lake District hills. This local variation may have a profound effect on the flora and vegetation.

What data we have for temperature on British mountains indicate that the conditions are very severe. Thus the long series of meteorological data, available from the station kept on the summit of Ben Nevis (4406ft.) from 1884 to 1903, shows that the mean monthly temperature at that summit is at or below freezing point for eight months of the year; and although this is the extreme case of altitude, mean monthly temperatures at 2000 ft. do not much exceed 50°F. In fact *low* summer temperatures, due to cloud and high winds, are characteristic of the British mountain climate and in rather sharp contrast to the mountains of more continental regions, such as the Alps. This must have a great effect upon the growth of mountain plants, whose available growing season is in any case severely restricted. Professor Gordon Manley, whose excellent book in the *New Naturalist* series (Manley, 1952) should be consulted for a detailed exposition of all aspects of the British climate, has stressed the significance of the rather small mean temperature variation throughout the year for the altitudinal limits of vegetation on British mountains. He points out that a small change of altitude in this country will produce a very much greater reduction in length of available growing season for mountain plants than a similar change in the Alps; and he correlates with this the low natural tree-line on British mountains, which is surprisingly low to visitors from Scandinavia, for example, who are familiar with coniferous forests ascending to above 2,500 feet in the Norwegian mountains.

Snow-lie is another factor of very great importance in mountain regions. No British mountain at present carries really permanent

B. Thrush

Plate III. A Welsh cwm, Glaslyn, with Snowdon behind

R. Atkinson

Plate IV. Wood sorrel, *Oxalis acetosella*, at 2000 ft. on Penyghent, Yorkshire snow-fields—the snow-patches of Ben Nevis and Braeriach (Cairngorms) come nearest to permanency—but a relatively slight general worsening of conditions could in fact bring the highest summits within the region of permanent snow. Manley has calculated that in the Ben Nevis region the permanent snow-line in the present-day climate would be at about 5,300 feet—that is, if the highest mountain in the British Isles were less than 1000 feet higher, it might well carry small glaciers and permanent snow-fields. Duration of snow-lie depends mostly on topography and exposure and varies enormously over very small areas, the exposed summit ridge, for instance, being swept bare of snow throughout the winter, whilst the adjacent corrie will accumulate deep snow-drifts. Snow-cover is significant to the mountain plant in affording winter protection, but, if persisting late, in severely reducing the growing season; and the distribution of many high mountain plants can be related more or less obviously to snow-cover (Pl. 1, p. 36).

For convenience, we can discuss separately the soil and climate factors; but we must realise that the former is intimately dependent upon the latter. The soil characteristic of upland Britain is an acid one, with a greater or smaller development of peat on the surface. Such peaty soils, and even blanket bog peat itself, can develop in regions of cool summers and heavy rainfall, whatever the nature of the parent rock, provided the ground is level or only gently sloping so that the peat can accumulate. In this way the greater part of the British mountain area is covered by acid-loving vegetation in which, as we shall see, rather few different species may be abundant over enormous stretches of land. Furthermore the processes which cause the leaching of bases from the soil and the acidification of the surface layers are aided by the growth of these familiar moorland and bog plants, the remains of which gradually accumulate as a surface layer of peat. The process is then a cumulative one, involving an interaction of climate with soil, vegetation and the complex of soil organisms responsible for decay; it is not *simply* reversible, although on many British mountains thick layers of blanket bog are apparently rapidly degenerating at the present time (see p. 35).

We can perhaps best picture the general relations of mountain vegetation to the conditions of soil and climate in some such way as this. Since the emergence of the British Isles from the Ice Age, the one-way processes of soil leaching, with subsequent podsol or peat-bog formation, have been going on at very different rates in different parts of the mountain areas, most rapidly where drainage and erosion was poor and rainfall heavy, so that acid peat quickly accumulated, and most slowly, if at all, where a combination of topography and rock type has meant that erosion has prevented any proper stable soil formation at all, and a copious water-supply supplies the necessary soluble bases to the plants growing there. Between these two extremes —well represented in Scotland, for example, by the Moor of Rannoch on the one hand and the steep slopes of a mica-schist mountain such as Ben Lui on the other—there are, of course, all gradations, and many more complications.

Undoubtedly the most important of these factors complicating the picture is the one generally called the 'biotic' factor, under which we may group all human and animal interference such as burning and grazing. Professor Pearsall's book in this series (1950) has dealt so admirably with the evidence of vegetational change in British mountains due to the influence of man and his grazing animals, that we shall be content here with a mere summary.

Grazing herds, particularly of red deer, were certainly a factor in the natural vegetation of Britain in the days before the main effects of human interference began to appear, when men lived by hunting; but there is evidence that the red deer was then naturally (as it still is on the Continent) an inhabitant of open woodland. The present-day association of red deer with the Scottish mountains is due largely to the wholesale destruction of forest by man, and, more recently, to the Victorian establishment of protected 'deer-forests' in the mountain areas. Forest destruction in upland (as in lowland) Britain was in part direct—for fuel and, by fire, to gain agricultural land—but also indirect, through the vast increase of man's grazing animals, and particularly sheep.

There seems little doubt that the remarkable development of sheep-rearing, which began on a large scale, particularly in Northern England, after the Norman Conquest, has been responsible, more than any other single factor, for imposing a stamp of uniformity upon large areas of mountain vegetation. The Pennines, the Lake District and the Welsh mountains are largely clothed by rather uniform grassland which has been sheep-walk for several hundred years. In the Highlands, forest destruction and the spread of sheep-rearing are both much more recent, and the latter has been in any case greatly restricted by the establishment of 'deer forest'. Another relatively recent development which has caused great modification of upland vegetation, particularly in the Pennines and parts of Scotland, is the establishment of 'grouse moor', with a policy of regular burning to encourage the growth of more or less pure stands of heather (*Calluna vulgaris*).

It is difficult to overestimate the effect of these kinds of interference, particularly sheep-grazing, on the British mountain flora. The immediate reaction of any Scandinavian botanist visiting British mountains is surprise at the extent of the uniformity and floristic dullness. Whole communities of plants, containing many characteristic mountain species, are in Britain represented by the merest fragments clinging to relatively inaccessible rock-ledges where the almost ubiquitous sheep cannot penetrate. This is particularly obvious, of course, on mountains where the rock is relatively rich in bases and on which, therefore, sheep-walk grassland is most easily and profitably established (see Ch. 9, pp. 157-8). From the point of view of the mountain flora, sheep and deer-grazing have the effect of greatly accentuating the contrast—which would be present to some extent for reasons of soil formation which we have already discussed—between the main, floristically dull, mountain communities, and those of unstable rock ledges on cliffs and steep slopes.

We are now in a position to sketch the general distribution of vegetation on British mountains, mentioning the characteristic dominant species. The lower flanks of the mountain up to 1500 ft. (or higher, depending upon conditions of exposure) presumably formerly carried woodland, either oak, birch, or, in parts of the Scottish Highlands, pine. This rarely survives today, even in a degenerate form, and has usually been replaced by some kind of grassland, moorland or bog, which can be looked upon as downward extension of communities naturally present to a more limited extent on the mountain above. The balance between grassland and moorland types depends in part upon the grazing and treatment of the area, but also on the occurrence of 'flushed' soils (through which relatively base-rich water seeps) and 'leached' soils respectively.

The extreme type of 'moorland' mountain is well represented by the Cairngorms (described in more detail in Chapter 6), where the parent rock is a coarse porous acid granite, and leached soils readily develop. Here a rough zonation of communities can be traced from the (still surviving) forest of pine and birch, through a zone of dominant

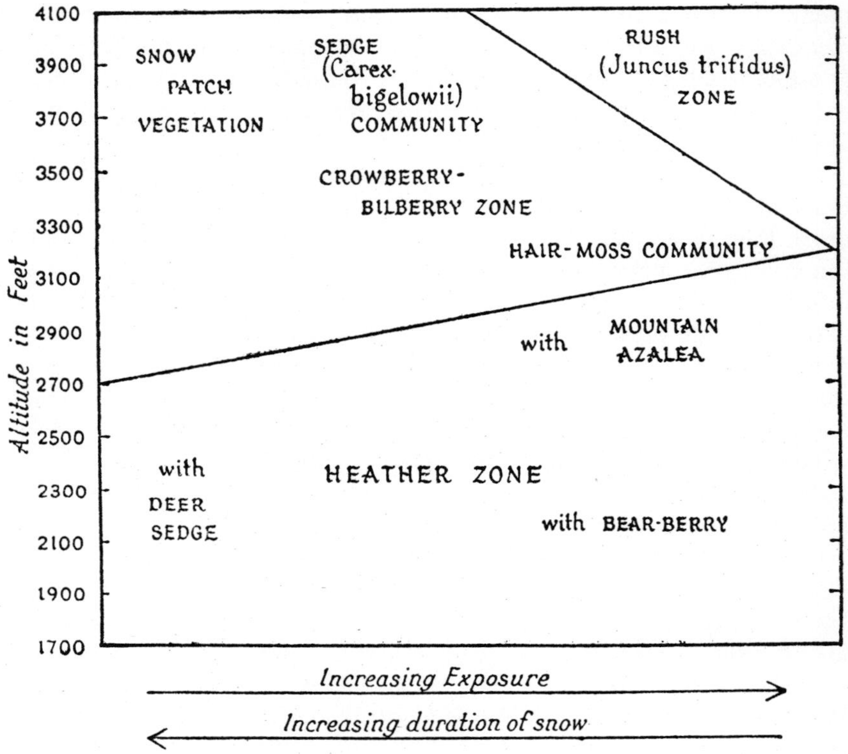

FIG 2

Diagrammatic representation of the interaction of altitude, exposure and snow-cover on the distribution of the chief plant communities in the Cairngorms (after Watt and Jones, 1948, *J. Ecol. 36*, 247)

heather (*Calluna*), to a zone where crowberry (*Empetrum*) and bilberry (*Vaccinium myrtillus*) are the main dominants. Above this zone is the summit vegetation, a sparse, open community of the moss *Rhacomitrium* and a very few flowering plants; and in patches in the dwarf shrub zones of crowberry-bilberry and heather are flushed grasslands mostly of mat grass (*Nardus*) or tufted hair-grass (*Deschampsia caespitosa*) (cf. Ch. 6 and diagram of zonation, fig. 2). Grazing (by deer) on such a mountain is relatively slight. In contrast, the familiar south slope of Ben Lawers, for example, shows a rather uniform sheep-walk grassland (with indications of the former extension of deciduous woodland in the shape of surviving relics by the steep-sided burns) which grades insensibly into the natural mountain grassland of the higher ridges. Dwarf shrub communities dominated by heather, bilberry and crowberry are poorly developed, because of the high base status of the soil (cf. Ch. 9).

On all types of mountain, whatever the parent rock, acid peat bog may be found where the slope is insufficient for free drainage. Bog vegetation, moreover, may occur at any height on the mountains, though the most extensive areas are naturally not at great altitudes, but occur, for example, on the ill-drained floors of glaciated valleys, or over gently undulating ice-scraped terrain as on the vast Rannoch Moor. There is also a tendency, as already mentioned, for peat bog to cover even appreciable slopes in regions of high rainfall, constant high humidity and cool summers; and we therefore find in Ireland and the Western Highlands vast areas of upland country covered with such 'blanket bog' or a mixed community somewhat intermediate between moorland and bog.

The active, growing bog, in which peat is accumulating, is characteristically dominated by species of bog-moss (*Sphagnum*). As in moorland, the flowering plants common in bog habitats are relatively few in number; there is the purple moor-grass (*Molinia*), the deer-sedge (*Trichophorum caespitosum*) and the cotton-grasses (*Eriophorum* spp.), the cross-leaved heath (*Erica tetralix*) and, of course, heather (*Calluna*) itself. A very constant adornment of upland bogs is the bog asphodel (*Narthecium ossifragum*), whose racemes of orange flowers are often produced in great abundance in July, and the insectivorous sundews, *Drosera rotundifolia* and *D. anglica*.

In general, however, actively-growing *Sphagnum* bogs are rare on mountains. At the present day the vegetation of areas covered with a thick peat layer is rarely dominated by bog-moss—although an investigation of the peat generally reveals that, at some previous age in the history of the bog, bog-moss was the dominant plant. Indeed, in some cases, the peat-deposits themselves are quite rapidly eroding away and over considerable areas there may be only a very incomplete cover of vegetation over bare peat. This is particularly striking in the Southern Pennines, where the flat tops of Kinder Scout and Bleaklow are covered by cotton-grass moor intricately dissected by rapidly-eroding drainage channels.

There seems good reason to believe—the evidence is admirably set out in Professor Pearsall's book in this series—that most bog or mixed moorland types in British mountains have been modified from a state more nearly approaching *Sphagnum* bog by precisely the same kind of interference factors—namely drainage, burning and grazing—as have greatly modified the moorland and grassland communities. There is also evidence, as we shall see in Chapter 4, in favour of a climatic change since the age when many of the thick blanket bog peats were laid down, a change which may on the whole have worked in the same direction as these interference factors, causing the cessation of active bog growth and, eventually, wasting of the peat layer.

Bog, moorland and grassland—these are the three vegetation types which cover all but a small fraction of the mountain area of Britain. What of the remaining fraction? Here classification is far more difficult; but as we have already seen, it is in the fragmentary, open communities on mountains that we find the vast majority of members of the British mountain flora, and in fact it is with such plants and such habitats that the main body of this book is concerned.

The exposed summit ridges afford an obvious example of an open plant community not classifiable as bog, moor or grassland. Here the moss *Rhacomitrium lanuginosum* is often the most abundant, even locally dominant, plant, and with it usually occur abundantly the sedge *Carex bigelowii*, one or two grasses, and dwarfed plants of bilberry (*Vaccinium myrtillus*). A good many other plants may be found in such exposed *Rhacomitrium* communities, and we shall mention particular examples in the following chapters. Scree slopes afford a somewhat similar habitat on many mountains, though here, of course, the sheer mechanical instability of the substratum makes colonisation by plants a difficult achievement. *Rhacomitrium* is again, however, a frequent effective coloniser, with individual higher plants of generally larger stature than the summit ones, such as the grass *Deschampsia caespitosa* and the parsley fern, *Cryptogramma crispa*, which root firmly into the soil between the loose rocks. Stabilised screes will eventually become effectively colonised by the dominant moorland plants such as heather.

Mountain cliffs and gullies afford one of the most extreme habitats, and, as we shall see in later chapters, most of the rarer arctic-alpine species are concentrated into such places.

W. G. Smith, in his classical account of mountain vegetations (in Tansley, 1911), subdivided mountain rock plants or 'chomophytes' according to their type of habitat in the following way. He distin-

M. E. D. Poore

Plate 1. Exposed summit ridge in Winter. Breadalbane Range.

B. Perkins

Plate 2. Rose-root, *Sedum rosea*, on cliff. Lawers range

guished first the plants forming very open communities on exposed rock-faces. The typical growth-form is a compact rosette (e.g. *Saxifraga nivalis*), mat (e.g. *Dryas*) or cushion (e.g. moss campion, *Silene acaulis*) with a single rootstock penetrating deeply into a fissure; here in this open habitat the smaller and rarer 'alpines' are to be found. The type of rock is of importance in two ways; firstly, in the soluble mineral salts it supplies to the roots of the particular plants, and secondly, in its ease of weathering to enable the roots to penetrate cracks. Thus, as we shall have occasion to appreciate many times throughout the book, the soft, calcareous rock which is rich in bases is richest also in diversity of mountain plants; next come the harder rocks which are base-rich (e.g. basalt) and soft rocks rather poorer in bases; and finally those rocks, of which much of the Scottish Highlands is composed, which are both hard and base-poor. On cliffs formed of such hard acid rocks it is not unusual to find the vascular flora consisting of two or three species only; the starry saxifrage (*S. stellaris* Pl. 4, p. 45) rarely fails us, even on the 'dullest' Highland cliff, and with it one can usually find tussocks of an often viviparous alpine form of the tufted hair grass (*Deschampsia caespitosa*)—but that may be all.

Smith's second group of cliff plants are those of sheltered ledges, where a certain amount of stability has allowed a depth of moist humus to accumulate. These cliff ledge communities are often most luxuriant, particularly of course on the mica-schist mountains. The plants of these ledges on mountain corries are partly 'true' mountain species such as the rose-root (*Sedum rosea,* Pl. 2, p. 37) or the sedge *Carex atrata,* partly plants which are commoner in northern Europe but not strictly limited to mountains, such as the globe-flower (*Trollius europaeus*) or the cranesbill, *Geranium silvaticum,* and partly mountain forms of widespread lowland plants such as the sorrel (*Rumex acetosa*). There is good reason to believe that, on the richer mountains, these 'herb-rich' communities would spread from their very restricted ledges to occupy some of the surrounding mountain grassland zone were it not for the intensive grazing activities of the sheep.

The third group proposed by Smith are the 'shade chomophytes', growing in deep crevices or in hollows between boulders, in situations where snow might persist until late in the season. Such plants may be found amongst the stabilised block scree at the base of many mountain corrie cliffs, or in shady ravines up the cliff face itself. This habitat is not of course clearly separable from the last, but several plants are particularly characteristic of it, notably the bladder-ferns (*Cystopteris fragilis* and the beautiful and rare *C. montana*) and holly fern (*Polystichum lonchitis*). Among lowland plants common in such situations we may especially mention wood-sorrel (*Oxalis acetosella*, see Pl. IV, p. 31).

Lastly Smith distinguished the streamside rock habitats; but these could equally well be considered under streams and springs, which we mention briefly below.

In places where snow lies late in the summer the vegetation is very characteristic, and again the communities, at least of the inner zone of the snow patch, are 'open' ones in which the higher plants occur as scattered clumps or individuals. No British mountain plant is confined to such habitats, but several occur frequently in these places, and their presence together, as scattered individuals in a very characteristic carpet of mosses and liverworts, is usually a safe indication of late snow-lie; among them are starry saxifrage, upright clubmoss (*Lycopodium selago*), alpine lady's mantle (*Alchemilla alpina*), least cudweed (*Gnaphalium supinum*, Pl. 3, p. 44) and *Sibbaldia procumbens* (Pl. XVII, p. 160). Around the moss carpet with the scattered higher plants may be a zone of the sedge *Carex bigelowii* and the least willow, *Salix herbacea* (Pl. 5, p. 64).

Springs and streams on mountains also provide small specialised habitats for plants; but again, as with the snow-patch, the dominant plants are often mosses. Flowering plants which are very commonly found in cushions of moss in these places include the inconspicuous water blinks (*Montia fontana*), the alpine willow-herb (*Epilobium anagallidifolium*) and saxifrages, particularly *S. stellaris*. Mountain lakes and tarns have a very poor flora generally. No flowering plant occurs exclusively in such habitats in Britain, although the quillworts (*Isoetes* spp.), the water-lobelia (*Lobelia dortmanna*), and the tiny relative of the plantain, *Littorella uniflora*, all occur frequently in stony shallow lakes at high altitudes. Pools and small lakes with peaty margins and acid water are particularly poor in aquatic flowering plants.

Such a delimitation of the main plant communities, each with its characteristic set of plants, should not be allowed to obscure the fact that, in detail, mountain vegetation often shows an intricate mosaic or some more regular alternation. Thus the open community of an exposed summit ridge will contain smaller areas of more closed vegetation, developed where shelter on the lee side of a large boulder, or under a low cliff, permits the effective growth of plants more characteristic of the mountain slopes. In this way crowberry and bilberry, for example, may form locally a closed community in the open *Rhacomitrium* zone; or, in cases where local shelter allows the accumulation of deep snow, a whole series of more or less definable communities may be present in zones, reflecting the importance of the factors of snow-cover and available moisture, in addition to the complex ones of exposure.

A common phenomenon on mountains is the arrangement of vegetation in waves or bands, often running like contours on a slope or across a gently-sloping ridge. In many cases this appearance can be shown to be associated with a terracing of the ground into a series of wide, shallow steps, the different vegetation strips being arranged in the corresponding position on each terrace. The cause of these and other wave and terrace patterns is still little investigated on British mountains, but they seem to be basically due to what is called 'solifluction'—the tendency of the upper soil layer, particularly when affected by alternate freezing and thawing, to 'flow' easily on even quite gentle slopes. To use a very crude analogy, the 'skin' of soil flows down in waves like treacle poured out of a jar. In addition, curious effects involving sorting of the larger rock fragments into rows or strips are of great importance on mountains, and terracing may partly involve sorting of this kind. In the Arctic the remarkable 'stone polygons' are produced by similar effects on more or less level ground, but these seem rare on British mountains.

In these examples of small-scale mosaic or zonation, the vegetation arranges itself, as it were, with respect to certain local features which are themselves constantly subject to change—the rock is shattered by frost, the soil flow subtly alters the terrace pattern. There is, then, a dynamic equilibrium, in which the mountain plants, by vegetative spread or by the establishment of seedlings, are constantly attempting the colonisation of bare rock and soil. In cases where they are being successful we may speak of a vegetation succession, consisting of a series of plants or communities of plants each achieving, in time, dominance over the previous one, until a more or less stable end-point is reached. Such successions would normally start with the lower plants (lichens, liverworts and mosses) and follow with one or more stages involving higher plants. It is quite possible to view the open communities of exposed ridges, cliffs, or snow-patches as early stages in the natural succession which is being prevented from advancing by extreme environmental conditions; if the particular factor, such as wind exposure or soil instability, ceases to operate in such an extreme manner, the succession can move forwards and the higher plants establish themselves.

In all mountain vegetation, then, but particularly that of unstable cliffs and screes, we are aware of the impermanence of the plant cover and the importance of the study of change. Some of the long-term results of this will be set out in the next chapters, where we consider the nature and origin of the present-day mountain flora; but no botanist who has scrambled around mountain cliffs can be under any illusion as to the importance of short-term change, often catastrophic change, to the mountain plants themselves. Erosion by wind, and by water in its various guises, is a very powerful factor constantly at work on the mountain habitats, creating new ground for colonisation and eliminating older, more or less established communities. Those plants which remain at the present day clinging on to life in these unstable surroundings must have been selected gradually throughout the years from a much larger number whose habit of growth and reproductive powers are unsuitable for survival in such specialised conditions.

CHAPTER 3

NATURE OF THE MOUNTAIN FLORA (M.W.)

THE BRITISH mountain flora contains some 130 different species of ferns and flowering plants (excluding 'critical' groups of micro-species) which are principally or exclusively at home above 2000 ft. (see Appendix I). As we have already seen in the previous chapter, many of these are rare or local plants, playing a relatively insignificant rôle in the mountain vegetation as a whole. The dominants in most 'closed' communities on mountains are either grasses, or dwarf shrubs mostly of the heath family, *Ericaceae*, all familiar in many lowland regions also; it is obviously of first importance to know these very abundant moorland and mountain plants if one is to learn the whole mountain flora. The most important are :—

(1) GRASSES

sheep's fescue, *Festuca ovina* (often on mountains represented by a viviparous plant which has been treated as a distinct species, *F. vivipara* (see p. 53))
bent-grasses, *Agrostis* spp.
mat-grass, *Nardus stricta*
purple moor-grass, *Molinia caerulea*
tufted hair-grass, *Deschampsia caespitosa*
wavy hair-grass, *Deschampsia flexuosa*
['deer-grass' or deer-sedge, *Trichophorum caespitosum*; this is not a grass, but a member of the sedge family, *Cyperaceae*].

(2) DWARF SHRUBS

heather, *Calluna vulgaris*
bilberry (blaeberry, whortleberry), *Vaccinium myrtillus*
cowberry, *Vaccinium vitis-idaea*
crowberry, *Empetrum nigrum* and *E. hermaphroditum* (see p. 54).

In addition to these, there are, of course, a considerable number of plants which, though not by any means characteristic of mountains, occur more or less frequently above 2000 ft. Often these mountain representatives of familiar species differ somewhat in habit of growth and other ways from their lowland relatives, and they have in certain cases been given different names—for example, the dwarf golden rod which occurs in exposed places on mountains has been called *Solidago cambrica* and treated by some authorities as a different species from the lowland *S. virgaurea*. This whole problem of the origin of mountain 'races' is discussed in more detail on p. 77. Appendix I gives a list of typically lowland species occurring not infrequently above 2000 ft. in the British Isles; it is useful to have such a list for reference when trying to identify mountain plants.

An interesting feature of British mountain plants is that systematically they are remarkably diverse. Almost all the large families of flowering plants in the British flora have at least one truly mountain species—perhaps the carrot family *Umbelliferae* and the mint family *Labiatae* are the only important exceptions, and in each of these cases we have species which are mountain plants in Europe, though they hardly qualify as such in their British habitats. Further, we often find that the mountain species occur as representatives of genera the majority of whose species are more familiar lowland ones; for example, there are mountain willow-herbs (*Epilobium*), speedwells (*Veronica*) and sedges (*Carex*). Perhaps the outstanding example of a group whose British representatives are predominantly mountain plants are the saxifrages (*Saxifraga*); and even here we have only a small selection of the wealth of beautiful species to be found on mountains on the European continent. We shall discuss the possible reasons for this when we consider later the evidence of the origin of the mountain flora in the British Isles.

Growth-form

If the systematic relationships between mountain plants are often rather distant, the resemblance in general appearance and habit of growth is, on the other hand, quite striking. This is a phenomenon found in many different kinds of specialised habitat, in which the growth-form apparently most suited to the conditions is possessed by a variety of unrelated plants. On mountains, many plants have the 'cushion' type of growth, with numerous short stems closely crowded together, in an extreme form making a hemispherical cushion. Such a growth form is characteristic of many mosses, and flowering plants such as moss campion (*Silene acaulis*) and mossy cyphel (*Cherleria sedoides*) have common names which draw attention to this feature. Cushion plants show the extreme type of adaptation to the mountain habitat; but many mountain plants (e.g. the least willow, *Salix herbacea*) possess a more or less prostrate branching and rooting habit of growth, and there is of course no clear demarcation between the two types. A third type is the rosette-forming plant, well illustrated by the saxifrages *S. stellaris* and *nivalis*, in which there is a very compact cluster of radical leaves, from the centre of which arises the flowering stem.

All these growth-forms are obviously well adapted to conditions of severe exposure, such as are to be found on the windswept summits of Scottish mountains; and under such inhospitable conditions the few flowering plants which survive are mostly of the cushion, mat or tussock types. All mountain habitats, however, are not subject to extreme exposure, so that we would expect that there are other factors which also favour the dwarf low-creeping or branching types of growth. We can get some indication of how one of these factors operates by looking at mountain plants growing in a habitat extremely different from the exposed, eroded summit, namely the 'snow-patch' where, because of the presence of a sheltered and usually shaded hollow, snow accumulates and lies late into the summer. Again very low-growing flowering plants are characteristic, though they are mostly rather loose mat-formers such as alpine lady's mantle (*Alchemilla alpina*) and least willow (*Salix herbacea*), rather than cushion types. In such a habitat, the over-riding factor is the curtailment of the growing season by the persistent snow-cover, and the successful plants are on the whole those which can grow rapidly as soon as the snow has gone.

The snow-patch is in this respect only an extreme case of a general condition on mountains. The effective growing season is normally short, and may in exceptional years be practically non-existent. We would not therefore expect to find many plants on mountains which need a long period of vegetative growth before flowering; they must as it were concentrate all their efforts into a short, sharp burst of growth. This phenomenon of sudden and early flowering is very familiar to gardeners, who grow a wide variety of the most beautiful 'alpines', mostly species at home in the Pyrenees or the Alps, in pans or on rock gardens.

There is another way of looking at the problem of the dwarf growth of mountain plants. Light has the effect of retarding the growth in length of shoots, and plants growing under conditions of high and continuous illumination will be stockier, with shorter internodes, than similar ones grown under low light intensity. We speak of this effect of pale, 'leggy' growth due to low illumination as 'etiolation'; it is familiar to gardeners in the results of chitting potatoes or forcing rhubarb. Now the illumination on high mountains can be very intense, and some of the familiar dwarfness of mountain plants may be therefore a direct effect. It would certainly be an important effect in the Arctic during the continuous light of the short summer; and, as we shall see later, most of our mountain plants are also widespread in the Arctic regions.

A successful mountain plant is likely, then, to possess a growth-form which enables it to grow and flower rapidly, utilising food reserves accumulated in the previous season; it will be, in other words, a *perennial* plant. There are two common types of perennial mountain plants; the *hemicryptophytes*, or ordinary herbaceous perennials whose over-wintering buds are protected, usually surrounded by the previous year's leaf remains, at or just below the soil surface; and the *chamaephytes,* mostly dwarf shrubs on whose persistent twigs the over-wintering buds are situated above the soil level. The hemicryptophytes include the rosette plants, such as the starry saxifrage, and the tussock or mat-forming herbs, such as the grasses, rushes and sedges. In addition to the dwarf shrubs (such as bilberry and crowberry) the chamaephytes include the woody mat-forming plants such as mountain avens (*Dryas octopetala*) and the more or less herbaceous cushion-formers such as the moss campion. Many mountain plants show remarkable speed in flowering on emergence from snow ; the flower-buds are often fully formed the previous autumn and remain dormant through the long winter.

It is interesting and significant that there are two types of perennial plant rare or absent in the mountain flora. One is the *phanerophyte* (shrub or tree with resting buds more than 25 cm. above the soil surface), whose absence is understandable in terms of exposure effects, particularly exposure of the resting buds unprotected by a winter snow-cover. The other is the *geophyte,* with resting buds buried in the

B. Perkins

Plate 3. Least cudweed, *Gnaphalium supinum.* Ben Lawers

J. Markham

Plate 4. Starry saxifrage, *Saxifraga stellaris.* Cairngorms soil, usually associated with a fleshy food-storing rhizome, bulb or corm. Many familiar lowland and garden plants are geophytes (e.g. bluebell, *Endymion nonscriptus*) and it is perhaps not immediately obvious why this type of plant should be very rare on mountains. The explanation seems to be that speedy growth as soon as the surface soil conditions are suitable is not possible for a plant whose resting buds are buried in a soil layer which may well remain frozen for a much longer period than the surface. Actually, the geophyte is most at home in the Mediterranean type of climate, where the unfavourable period for growth is not the cold winter, but the hot, arid summer. We have only a single representative in the mountain flora strictly defined—a slender bulbous plant of exceptional refinement and rarity, called *Lloydia serotina* (see pp. 12 and 104, and Pl. 9, p. 116).

Annual plants, which must come from seed every season, are also understandably rare in the mountain flora. The growing season is so short that only the most diminutive plant can expect to complete its life-cycle from seed every year; and what is more serious, a single catastrophically bad season may prevent the formation of ripe seed at all. Until recently, (neglecting the critical eyebrights and yellow rattles (see p. 74)), only one truly annual mountain plant could be claimed as British, the lovely snow gentian, *Gentiana nivalis* (see p. 171). The discovery and confirmation of the tiny arctic *Koenigia islandica* in Skye, however, (see pp. 135-6) has given us another example. Neither, of course, is a common mountain plant, though *Koenigia* is abundant over the small area in which it is known to occur.

Pollination

When we think of mountain plants, it is natural that those with large and brightly-coloured flowers should most readily come to mind —the 'alpines' of the rock gardener—and that we should therefore assume that insect-pollination is the commonest state. In the British mountain flora, however, wind-pollinated species make up quite a high proportion of the total—26%, in fact, of the plants listed in Appendix I. This figure compares with 16% calculated for the Alpine flora, and 32-39% for various arctic floras (Schroeter, 1926); the great difference between these last two is due very largely to the numerous species of arctic grasses and sedges—groups which are exclusively wind-pollinated. It is not surprising that in conditions of severe exposure, wind- rather than insect-pollination should predominate (as is suggested, for example, by the fact that just over 40% of the flowering plants recorded as growing above 2,500 ft. in the Clova mountains are wind-pollinated) (Willis and Burkill, 1901—see below). What is perhaps more surprising is the evidence which has accumulated that many mountain plants with insect-attracting devices (colour, scent, nectar etc.) are often if not exclusively self-pollinated. This is such a common situation that it deserves closer inspection.

A considerable literature is available about the pollination and general 'floral biology' of plants in the Alps, and also of what are often the same species in the arctic regions, but, so far as I am aware, there exists only one important contribution to our knowledge of pollination of British mountain plants. This study, undertaken by J. C. Willis and I. H. Burkill in the Clova mountains between the years 1894 and 1899, was fortunately, however, a remarkably thorough one, in which the co-operation of entomologists ensured accurate information on the species of insect visiting the flowers. Their findings were published in a number of papers between 1895 and 1908; it is a pity that they were unable to complete the work, as they had intended, by a final paper comparing in detail their observations with those available for the Alps, but even without this, the work is a most valuable contribution to the British literature in a surprisingly neglected field.

Let us take a few concrete cases from Willis and Burkill's data. They found, in the Clova mountain flora, only one species which was both structurally adapted and actually functioning as a butterfly-pollinated plant; this was the moss campion, *Silene acaulis,* for which 27 out of a total of 50 recorded insect visits were by Lepidoptera. They further noted that three types of moss campion plant occurred, bearing all male, all female and hermaphrodite flowers respectively, and that the last-named type was the commonest. In the Alps this species is also pollinated by butterflies, but the separation of the sexes is there much more complete, and therefore cross-pollination is more or less obligatory. In Greenland, on the other hand, the nectar of the moss campion flowers is apparently a good deal more accessible, and bees as well as butterflies pollinate (Pl. 7, p. 80).

As an example of a bee-pollinated flower observed by Willis and Burkill we may take the bearberry (*Arctostaphylos uva-ursi*). This is strongly scented, and commonly and apparently (in Clova) almost exclusively visited by species of *Bombus*, especially *B. lapponicus*, though, in parts of the Highlands, the arctic moth *Anarta cordigera*, whose larvae feed on the plant, is recorded as visiting the flowers (Ford, 1955, p. 153). It sets fruit well, and must be as effectively cross-pollinated in the Clova mountains as in the Alps. In Scandinavia and the Arctic it apparently shows in its flowers slight structural differences, as compared with the Alpine plant, which make self-pollination much more likely to occur. The other species of bearberry, *Arctous alpina,* however, has much less conspicuous whitish flowers which are apparently normally self-pollinated, both in the Alps and the Arctic; this plant was not observed by Willis and Burkill, but structurally the British plant agrees with the Continental one and is presumably self-pollinated.

Two Clova rarities were carefully observed by Willis and Burkill; these were the alpine catchfly (*Viscaria alpina*) and the yellow oxytropis (*O. campestris*). The catchfly, structurally a butterfly-flower, and functionally so in the Alps, was visited by very few insects on its high plateau in Clova (but see also p. 108)—actually the total recorded visits were one bee, three flies and four 'other insects' (presumably very small). Willis and Burkill conclude: 'Ultimate self-pollination is possible, and seems to take place, fruiting being regular'. The oxytropis fares better, apparently; it has a sweet smell and abundant honey, and attracted four bees, 15 flies and 78 'others'; but self-pollination is possible and the plant certainly fruits very freely.

As a final and extreme example of a structurally specialised flower observed by Willis and Burkill we can cite the butterwort (*Pinguicula vulgaris*). This proved to be habitually self-pollinated, and 'as neglected by insects about Clova as elsewhere'.

Willis and Burkill make a general comparison of the insect visitors to flowers which are structurally adapted for pollination by long-tongued insects only (butterflies and bees) in the Alps and in the Clova mountains. They show that whereas in these classes in the Alps there is a great preponderance of butterfly visitors, in Clova this is not so—there is actually an excess of visits by short-tongued flies over *any* other class, and butterfly visitors are distinctly rare. This very considerable difference must be a general one as between the Alpine and the northern floras and insect faunas, and is obviously at the root of the other general difference we have seen, namely a tendency for the same species to be more often and more easily self-pollinated in northern and particularly in Arctic regions than in the Alps. The conclusion is inescapable that mountain species with showy flowers adapted for specialised pollination by long-tongued insects seem to be, in that respect at least, relatively ill-adapted in northern Europe as compared with their Alpine range, where the pollination mechanism seems to work 'as it was designed to do'. The successful 'arctic' type is the simple open flower in which there is little or no concealment of nectar and self-pollination is possible or even probable (e.g. many small *Caryophyllaceae*). This type of flower Willis and Burkill found to be the commonest type among the 'alpines' of the Clova region. In the British flora, it seems that the number of mountain plants normally effectively cross-pollinated by insects is probably very low—perhaps the few dioecious species such as the rose-root (*Sedum rosea*), and a few such as the bearberry, which cater effectively for the humble-bee, are almost all we have. Many others, of course, may be occasionally cross-pollinated; but a surprisingly large number, one suspects, will prove to have the 'outward trappings' of a cross-pollination mechanism which does not in fact normally work.

This whole topic raises fascinating unsolved problems, and is moreover one in which the patient naturalist can usefully supply badly-needed information which might help to elucidate some at least of these. Take, for example, the case of *Bartsia alpina* (see pp. 164-6), described in British Floras as 'pollinated by humble-bees'. This is certainly true in the Alps, where Müller first described the pollination mechanism. The curious dull-purple corollas have a tube over half-an-inch long and the style matures early and protrudes from the flower well before the stamens are ripe. Humble-bees visiting the flowers for the nectar secreted at the base of the tube come into contact with the style and deposit foreign pollen on it. At a later stage, the ripe anthers protrude as the style withers. However, in other parts of Europe (in the Riesengebirge and the Norwegian mountains), the flower has been described as more or less homogamous (i.e. the style and stamens ripen at the same time), so that in these regions the flowers may often be self-pollinated. It would be of interest to know whether the British *Bartsia* shows protogyny and is in fact normally cross-pollinated by humble-bees, or whether it is homogamous and self-pollinated. So far as I am aware, no-one has ever observed it, yet thousands of botanists have visited the plant in one or other of its few British localities.

Before leaving the subject of insects and mountain plants, I would like to mention here a remarkable paper by that great Scots naturalist, Buchanan White, who wrote the Flora of Perthshire (published posthumously in 1898). Buchanan White was a first-class entomologist and botanist—a combination that seems, alas, to be much rarer today than it was in the last century—and was particularly interested in the relations between insects and plants. Nowhere is this more clearly revealed than in his paper on 'The Mountain Lepidoptera of Britain; their Distribution and its Causes' (White, 1879), in which he analysed the ranges, in Britain and abroad, of 15 mountain moths, and considered the historical and—to use a modern term unfamiliar to White—ecological factors in their distributions. He points out that though some feed in the larval stage on restricted mountain plants (one on the dwarf birch, *Betula nana*, another on saxifrages, principally *S. aizoides*), all seem to be much more restricted than their host plants, and suggests that climatic factors as well as food-plants are important in limiting the present-day range. The paper is also remarkable in its appreciation of the historic factor in interpreting plant and insect distributions, and will be alluded to later (p. 87).

Seed Production

For many mountain plants, as we have seen, seed production must be very variable from year to year. It is a common experience to see sterile rosettes of starry saxifrage, for example, newly uncovered in a snow-patch in late July. Such plants are quite unable to set ripe seed in the same year. Another species, the famous drooping saxifrage (*Saxifraga cernua*, Pl. 6, p. 65), rarely flowers in its two known localities in Scotland, and only reproduces vegetatively, chiefly by small red axillary bulbils, which drop off and over-winter like large 'seeds'. The fruit and seed of this saxifrage are indeed unknown, even in Arctic regions where the plant is common and flowers quite freely.

Detailed studies of seed production in mountain plants have not, so far as I am aware, ever been published in this country; but Professor Pearsall (1950, pp. 49-51) has given details of a very rewarding study of the growth of the moor-rush (*Juncus squarrosus*) at different altitudes on British mountains, from which he has shown that, whilst temperature factors retarding the development of flower and fruit normally effectively prevent any seed-production above about 2700 ft., the plant occurs up to at least a thousand feet higher, and can therefore grow healthily in a vegetative state well above its 'reproductive limit'. He suggests that transported seeds (the plant occurs particularly along sheep-tracks and trampled paths) are mainly responsible for the high-level plants; although after an exceptionally long, warm summer he found viable seeds at 3,400 ft. on Ben Wyvis. A similar relationship between external factors and seed production has been elucidated for another species occurring on mountains, the common wild thyme (*Thymus drucei*) (Pigott, 1954); and in general it seems likely that effective vegetative growth can take place in conditions of temperature, light, etc. well outside the range within which viable seed can be set.

An intriguing possibility suggests itself that, with certain rare mountain plants, conditions are no longer suitable in their habitats in this country for seed to be set in normal seasons, though in other parts of their range they may still reproduce effectively by seed. The recently-discovered *Artemisia norvegica* may well prove to be such; for the plant seems to flower rather sparingly, and there was no sign of developing fruit on it at the end of August, 1953. In the Dovre mountains of Norway, it sets seed regularly.

The two arctic species of *Rubus* are interesting in this connexion, as Salisbury (1942) has pointed out. *Rubus arcticus* itself (a 'doubtful' native of Britain, see p. 173) seems to fruit poorly or not at all over a good deal of the southern part of its range in Scandinavia, for example; and there is evidence that a similar situation is found with regard to the outlying southern stations of the cloudberry (*Rubus chamaemorus*) in Britain as on the Continent. Thus in many years and in many localities the fruit does not ripen in Germany (Hegi, Vol. 4, p. 766); and the isolated localities give the impression of being relict stations of an arctic species no longer growing under optimal conditions. In Britain, if one neglects a record of the occurrence of the plant in Pembrokeshire for which there appears to be no recent confirmation, the southernmost stations for the cloudberry are on the Berwyn range on the Montgomery, Merioneth and Denbigh borders. Here the plant is locally quite abundant in blanket bog above about 2,300 ft. As one might expect, such a distinctive plant with an edible fruit has quite a folk lore surrounding it, and there are interesting references to the cloudberry in local legend and history. Some of these references suggest that the plant flowers and fruits much less freely at the present day than it used to do; for example, the following extract from an article on the history of Llanrhaiadr-ym-Mochnant, a village at the foot of the Berwyn Range (Hancock, 1870).

"The following remarkable and interesting tradition [exists] respecting [St. Dogfan]; his name is associated with a certain plant growing on the Berwyn mountains called Mwyar Dogfan (Dogfan's berries) and Mwyar Berwyn (Berwyn berries) and the tradition current among the old Berwyn shepherds was that the wages of Dogfan Sant, in the year, for his spiritual ministries to his flock at Llanrhaiadr, was a quart of the Berwyn berries, and that the custom existed in the parish that whoever brought a quart of the ripe berries to the parson of the parish on the morning of the day of the saint's festival had his ecclesiastical payments remitted for the year."

The writer continues: "So scarce is the fruit in some years, owing to the elevated locality (above 2,000 ft. above the level of the sea) at which they grow, which subjects the flowers to nipping frosts in the flowering season, that a quart of the berries cannot be had throughout the range. The writer, in the year 1869, after the most diligent search, was only able to secure some half dozen berries. It is related that the late Sir Watkin Williams Wynne of Wynstay, the father of the present baronet, offered during some years as much as 5/- for every ripe berry brought to him. Our informant in this case could not tell the reason of Sir Watkin putting such a value on them. Edward Llwyd the great naturalist, writing to his friend, Mr. Lloyd of Ruthin, in 1683, begged him to get some of the berries and roots of this plant and send them to Oxford."

One might reasonably question Hancock's interpretation of the climatic effects responsible for bad fruiting; but, taken at its face value, the legend and history suggest that in the fifth century the cloudberry fruited rather more plentifully than in the nineteenth and, certainly, the mid-twentieth centuries.

The cloudberry is certainly a shy fruiter over much of the British Isles; but my limited observations on the Southern Pennines where the plant is abundant would suggest that the sterility of large patches is due in part, at any rate, to the greater frequency of the male plant. Quite large areas are occupied only by a single sex, which presumably indicates that vegetative spread by means of the slender rhizome is of great importance. The male and female plants may indeed be behaving to some extent as separate 'species' with different tolerances of climatic and soil conditions. They can exist independently and spread by vegetative means, and the situation resembles, though is less extreme than, that of the butterbur, *Petasites hybridus*, in which, over large areas of Britain, only the male plant is known.

It might prove possible in cases like that of the cloudberry to relate the decline in fruiting to climatic change over the country as a whole (cf. p. 88); although it is important to remember that the cloudberry is a clear case of a mountain plant with a very specialised habitat requirement. Other northern species, such as the cowberry (*Vaccinium vitis-idaea*) and the crowberry (*Empetrum*), will grow in a wide range of upland and mountain communities, from largely mineral acid soils to pure peat-bog. The cloudberry, however, is restricted, throughout its range in arctic and subarctic regions as on mountains, to peat-bog communities. This specialised soil requirement is undoubtedly one of the reasons for the extreme difficulty of cultivating it. Syme in his description of the plant in 'English Botany' (1863, Vol. 3, 159) makes the suggestion that, '. . . the gardener might find means to render this plant a valuable and useful addition to the kitchen garden by crossing the flowers with those of the bramble and the raspberry, and thus overcoming the tendency to flourish only away from cultivation'. A good deal of valuable genetical research and practical plant breeding has been devoted to the genus *Rubus* since this was written, but the cloudberry-blackberry hybrid is still an unfulfilled dream—and indeed is likely to remain so, for the cloudberry can only be very distantly related to the other species of *Rubus* in our flora.

Vegetative Spread, 'Vivipary' and Apomixis

Most mountain plants, we have seen, are perennials. With rare exceptions, therefore, they can spread very effectively by vegetative means, and may be more or less independent of reproduction by seed for their persistence, at least over quite long periods. Pollination, seed ripening and dispersal, and the establishment of seedlings are such critical stages in the complete life-history of a plant, that it is not surprising to find that, under the extreme conditions found on mountains, many plants show short-circuits of their complete life-cycle, spreading principally, or in some cases wholly, by means which do not involve all the hazards of sexual cross-fertilisation.

A single individual of the least willow (*Salix herbacea*, Pl. 5, p. 64) may cover several square feet with tiny rooting branches and twigs interlacing and largely buried in the stony soil. We do not know the age of such plants, or whether the individual can be said to die of old age at all. The rate of growth of the woody parts of these mountain plants must be extremely slow. Schroeter (1926) illustrates this by contrasting the cross-section of a 'trunk' of the dwarf azalea (*Loiseleuria procumbens*) which showed 55 annual rings, averaging less than 1/10 mm. in thickness, with a 6-year-old tropical tree whose annual rings averaged over 40 mm.

In addition to the simple vegetative spread shown by most mountain plants, there are several types of reproduction of a more specialised nature. One of these, bulbil production, we have already mentioned in the interesting case of *Saxifraga cernua,* in which no fruit and seed has ever been described; another familiar example of a mountain plant reproducing largely if not wholly by axillary bulbils is the common upright fir clubmoss, *Lycopodium selago.* In such cases the bulbils are produced as modified axillary branches associated with the leaves of the plant. Similar structures may, however, replace some or all of the ordinary flowers in the inflorescence. This condition is usually called 'vivipary' by botanists (although it should be noted that true vivipary —the germination of seeds while still attached to the parent plant— is an entirely different phenomenon), and is very well shown by the common mountain bistort, *Polygonum viviparum,* which produces a terminal spike usually an inch or two in length, in most of the lower part of which the flowers are replaced by purplish bulbils. In several mountain grasses, also, the inflorescence may be partly or wholly viviparous. Much the commonest is the fescue, *Festuca vivipara,* which largely replaces the ordinary species (*F. ovina*) on many Scottish mountains; there are also alpine meadow-grass (*Poa alpina*) which is usually viviparous in this country, although two stations are known where the 'normal' sexual form occurs (see p. 141), and the tufted hair-grass (*Deschampsia caespitosa*) and its close relative *D. alpina* which are often—the latter probably always—viviparous at high altitudes. Wycherley (1953a, and 1953b) has recently given much information on these viviparous mountain grasses.

These are by no means all the 'short-cut' devices shown by mountain plants. An important group appear to exhibit normal sexual seed production, and may even have showy flowers attractive to insects, but on investigation prove to be capable of setting some or all of their seed without any fertilisation. This phenomenon is known as *apomixis,* and a plant showing it is said to be *apomictic*[1]. Examples of apomictic groups of mountain plants are discussed later (pp. 75-6). Here it will suffice to say that such groups tend to be 'critical' in the British flora—that is, botanists find it difficult to decide the limits of the species, and have usually created a number of more or less closely similar 'micro-species' in an attempt to deal with the situation. Total apomixis, such as is shown by the dandelion (*Taraxacum*) and lady's mantle (*Alchemilla*), is a fairly easy phenomenon to demonstrate experimentally by preventing all pollination and yet achieving perfect seed production. But there are also a good many examples of partial apomixis in species which are principally sexually reproducing (e.g. the meadow buttercup (*Ranunculus acris*), cf. Turrill, 1948, p. 245) and these are naturally more difficult to work out. Even more difficult are those cases of 'hidden apomixis' or *pseudogamy*, in which pollination is necessary for seed production, but nevertheless there is no actual fertilisation involved. The alpine cinquefoil (*Potentilla crantzii*, Pl. 15, p. 190) affords an example of this; and, of course, there may well be others in the mountain flora of which we are as yet quite ignorant.

None of these specialised reproductive measures are, of course, confined to mountain plants, but it is true to say that the proportion of species relying largely or wholly on non-sexual methods of reproduction is much higher in the mountain flora than in any lowland one.

Another interesting point which has some bearing on this question is the high proportion of polyploid species among mountain plants—indeed among Northern plants in general in Europe compared with, say, those of the Mediterranean region (cf. Løve, 1953, and references given). Two examples may be cited. The moorland crowberry, *Empetrum nigrum,* is a diploid plant (2n = 26); its close mountain relative *E. hermaphroditum* (abundant on Scottish mountains, where it largely replaces *E. nigrum*) is tetraploid (2n = 52). The mountain saxifrages *S. caespitosa, S. cernua, S. nivalis* are all high polyploids. It is only fair to say, however, that it is by no means impossible to find examples of the opposite kind; thus, the Arctic form of *Vaccinium uliginosum* is a diploid, whereas the common North European (and the British?) plant is tetraploid; and, in the genus *Thalictrum,* our mountain

[1]Gustafsson (1946), the Swedish authority on the subject, includes *all* examples of vegetative reproduction under the term *apomixis,* and uses the special term *agamospermy* for the production of seed without fertilisation. Whilst there is much to be said logically in favour of this, I have thought it best to use apomixis in the restricted sense as familiar to most British botanists.

species, *T. alpinum*, has the diploid number, whilst the other two lowland species are polyploids.

It is certainly true that, in the majority of cases of pairs of closely-related species, the one diploid and the other polyploid (usually tetraploid), the polyploid shows a greater vegetative vigour and ability to persist and spread by producing adventitious roots. Familiar examples are the two species of bird's foot trefoil, the common tetraploid *Lotus corniculatus* and the more local, lowland, slender bird's foot trefoil, *Lotus tenuis*, which is diploid; here the tetraploid is the widespread, successful species occurring from sea-level to high altitudes on mountains, and spreading freely vegetatively. A further example is afforded by the two bitter-cresses, *Cardamine hirsuta* and *C. flexuosa*. The former is an annual, with no power of vegetative spread, and is diploid ($2n = 16$); the latter is usually biennial or short-lived perennial, with more vigorous stems, readily rooting at the nodes, and is tetraploid ($2n = 32$). Both these species, incidentally, are said to occur up to nearly 4,000 ft. on Scottish mountains. There are several possible ways of interpreting this evidence. We might suppose, for instance, that low temperatures actually induced the formation of polyploids more or less directly—an attractively simple idea, for which, unfortunately, little supporting evidence exists. A more complicated interpretation would be that polyploidy, by encouraging vegetative vigour in some way, is of advantage to arctic-alpine plants in general, and therefore is more widespread in that group. This seems more satisfactory, but still leaves a lot of questions unanswered. Further clues are found in our knowledge of the origin of the mountain flora, and we must leave the question at this point, to return to it later.

Types of Distribution

We have very few *endemic* species in the British Isles; that is, plants (or animals) which are not known at all outside these islands. The mountain flora, therefore, consists almost exclusively of plants which can be found elsewhere in Europe—often, indeed in much greater luxuriance and abundance than on British mountains. From the point of view of distribution outside the British Isles, we can, then, group our mountain plants into three main classes, viz. arctic, alpine, and arctic-alpine.

1. Arctic.

Certain British species occur in the Arctic regions and often also in the Scandinavian mountains, but are otherwise absent from the European mountain ranges.

These are:—

Alopecurus alpinus, alpine foxtail (not in Scandinavia)
Carex aquatilis }
C. rariflora } sedges
C. saxatilis }
Cerastium edmondstonii (*arcticum*), arctic mouse-ear chickweed
Chamaepericlymenum suecicum, dwarf cornel (southwards to N. Germany)
Deschampsia alpina
Diapensia lapponica
Draba rupestris (arctic complex including *D. norvegica* etc.)
Erigeron borealis, boreal fleabane
Juncus biglumis, two-flowered rush
Koenigia islandica
Luzula arcuata, curved woodrush
Minuartia rubella, alpine sandwort
Poa flexuosa
P. x jemtlandica
Rubus chamaemorus, cloudberry (southwards to S. Germany)
Sagina intermedia, lesser alpine pearlwort
Salix lanata, woolly willow
Saxifraga caespitosa, tufted saxifrage (fig. 3)
S. rivularis, brook saxifrage

It is interesting that this list contains only two common British mountain plants; the vast majority of them are both rare and local. In view of the fact that two of the most remarkable examples—*Diapensia* and *Koenigia*—have only been discovered in the last few years as British plants, it seems not improbable that other arctic species may still await discovery in the British Isles.

2. Alpine

British mountain plants found in the European mountains but not in the Arctic are even more restricted in number than the arctic group. They include:—

Alchemilla conjuncta, 'alpine lady's mantle'
Cherleria sedoides, mossy cyphel (fig. 4)

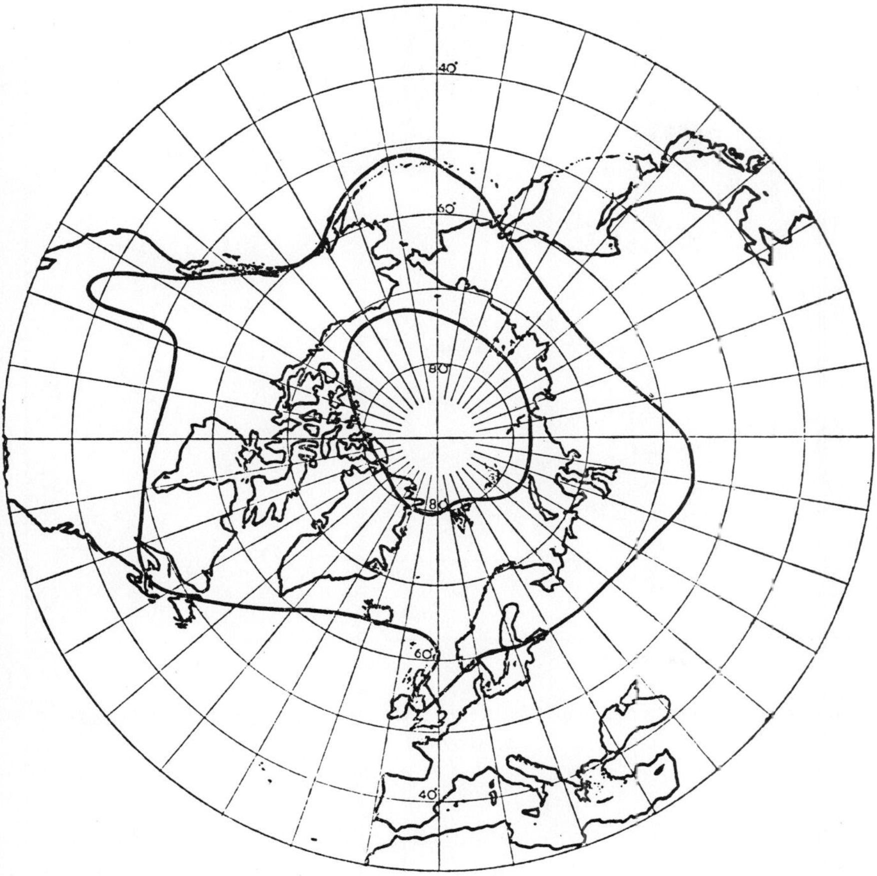

FIG. 3
World distribution of *Saxifraga caespitosa* (after Melderis, 1953, in Lousley (ed.), *Changing Flora of Britain,* p. 91)

Gentiana verna, spring gentian
Myosotis alpestris, alpine forget-me-not
Oxytropis halleri, purple oxytropis
Thlaspi alpestre agg., alpine penny-cress
Viola lutea, mountain pansy

Very few of these are really strictly mountain plants in the British Isles, and *Cherleria* is unique in the flora because it is the only member

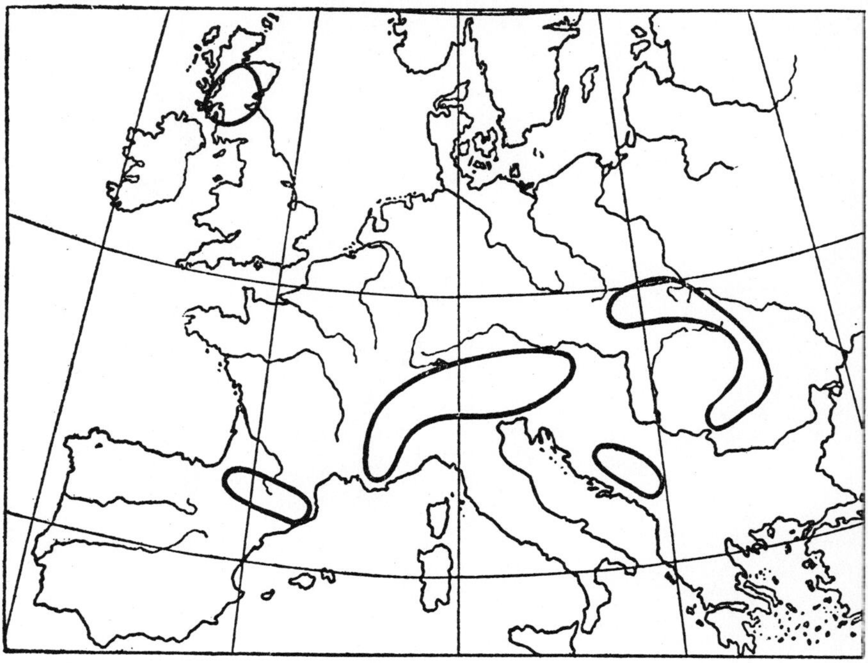

FIG. 4
World distribution of *Cherleria sedoides* (after Mattfeld, 1928-30, in *Pflanzenareale*, 2, Map 61)

of this group which is locally abundant quite high on mountains. The bird's eye primrose (*Primula farinosa*), which is quite common in damp grassy and marshy places on the Yorkshire limestone, is excluded from this list simply by adherence to the rule that a 'mountain plant' must have at least *one* British locality above 2,000 ft. Although, as we have already explained, we have not obeyed this rule strictly, it is convenient to use it in the case of species of 'alpine' distribution type, as it does afford a clear criterion for delineating an otherwise undefinable group. Anyone who has botanised in the Alps, where the bird's eye primrose occurs, widespread and often in great abundance on limestone up to over 8,000 ft., might well feel aggrieved by this decision; but we can only plead in defence that a line must be drawn somewhere, or alpine species such as the yellow whitlow-grass (*Draba aizoides*), which occur almost at sea-level on limestone in this country and are manifestly not mountain plants, would also claim a place in the list.

Such distributions also grade into those of some famous members of the Irish flora, which have generally been grouped in the 'Lusitanian' element, but which are better characterised as 'Pyrenean'. The group of 'Robertsonian' Saxifrages, to which the familiar garden 'London Pride' belongs, includes two native Irish plants, both of which occur at 2,000 ft. on the mountains in Kerry and Cork. One of these, *Saxifraga spathularis,* occurs elsewhere only on the N. Portuguese and N. Spanish mountains; the other, *S. hirsuta,* only in the Pyrenees and N. Spain. The Irish butterwort, *Pinguicula grandiflora,* which ascends the Kerry mountains to near 3,000 ft., has a distribution connecting these 'Lusitanian' or 'Pyrenean' types with the more familiar alpine ones, occurring in the N. Spanish mountains, the Pyrenees, the French Alps and the Jura. (This European distribution is incidentally closely paralleled by that of the Bristol rock-cress, *Arabis stricta,* which, however, in this country is almost restricted to limestone rocks in and near Bristol).

It is, of course, reasonable to expect that many species will on the whole occur at lower levels in the more northerly British Isles than, say, in the Pyrenees or Western Alps, and it is precisely this general tendency that makes the 'alpine' element in the British flora understandably difficult to define.

A small group of species which occur on European mountains, in Britain, and in Iceland is of particular phytogeographic interest. This group includes the blue moor-grass (*Sesleria caerulea*) and the slender bedstraw (*Galium pumilum*).

3. ARCTIC-ALPINE (fig. 5).

The vast majority of our mountain plants come in this category, occurring both in the Arctic and on some or all of the main European mountain ranges. This most striking feature of plant distribution early fired the imagination of botanists, as we shall see later; it is so generally true that in many ways a visit to the Swiss Alps or the Norwegian mountains affords a better opportunity of learning the majority of British mountain plants than would a visit to many Scottish mountains. Some of the commonest 'alpines' of gardeners, such as the purple saxifrage (*Saxifraga oppositifolia*) and mountain avens (*Dryas octopetala*), can be found, in slightly differing forms, throughout the

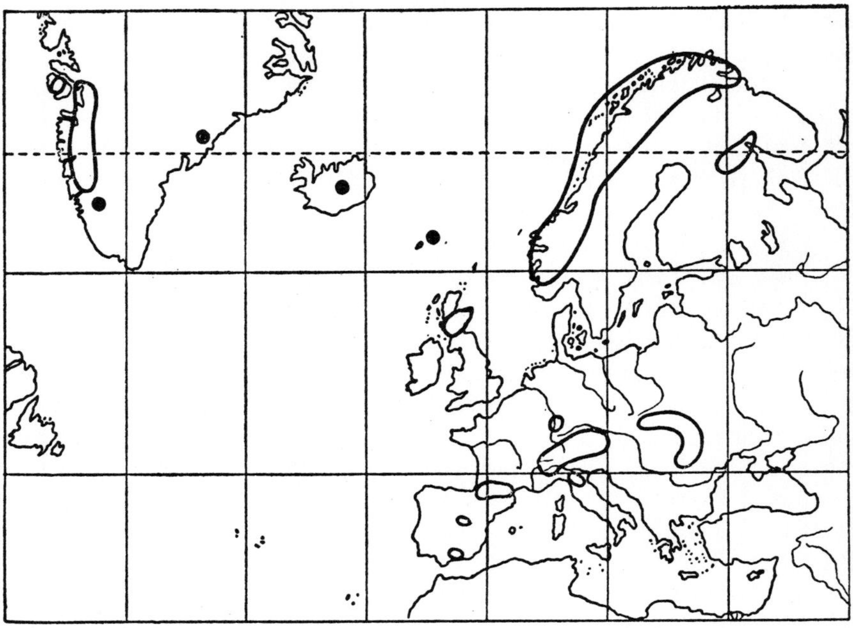

FIG. 5
World distribution of *Veronica fruticans* (after Huber, 1928-30, in *Pflanzenareale, 2,* Map 40)

Arctic regions, on the principal European mountain ranges, and also on mountains in North America. A few species (e.g. the sedge, *Carex lachenalii,* fig. 6) even occur on mountains in the Southern Hemisphere, apparently little if any different from the northern forms.

Within this broad category can be mentioned certain rather unusual types. Some few are absent from the Alps proper, while occurring elsewhere on European mountains—an interesting example is the tiny Teesdale rarity, *Minuartia stricta*, which occurs widely in the Arctic, but only otherwise in the Jura and the Bavarian Alps (fig. 7). Conversely, we have a few alpine species of somewhat restricted northern distribution, such as the yellow-flowered *Oxytropis campestris,* a familiar Swiss mountain plant very rare in Scotland. The beautiful blue sow-thistle (*Cicerbita alpina*) represents another possibility; it is a widespread European mountain plant which occurs abundantly in the Scandinavian mountains and reaches into the Arctic there. It is, however,

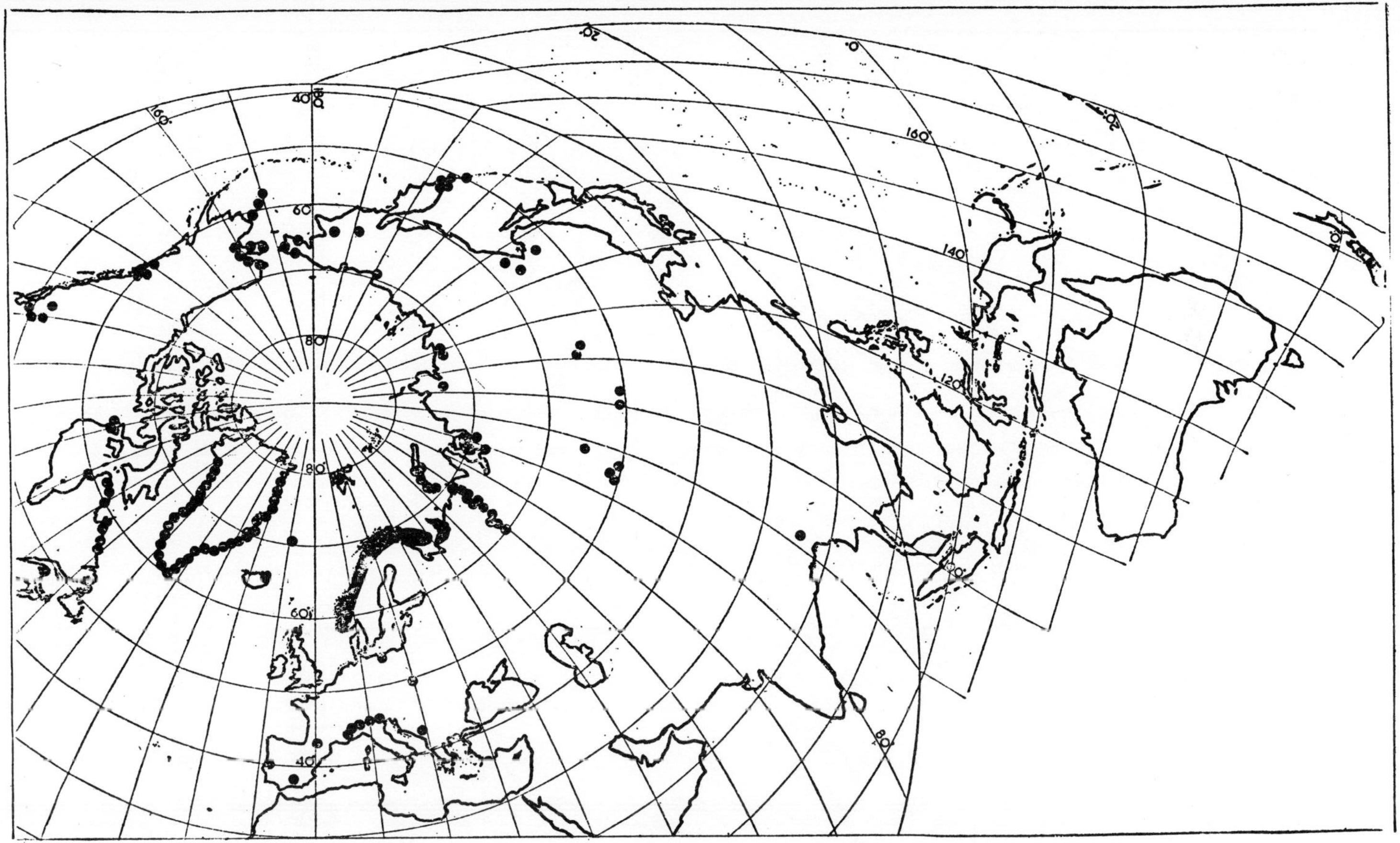

FIG. 6: World distribution of *Carex lachenalii* (after Du Rietz, 1940, *Acta Phytogeographica Suecica*, *13*, 216)

FIG. 7
World distribution of *Minuartia stricta* (after Mattfeld, 1928-30, in *Pflanzenareale*, 2, Map 56)

more strictly 'sub-montane' than 'montane', a plant of the upper coniferous and birch forests rather than of the open mountain.

4. SPECIAL TYPES OF DISTRIBUTION.

A few mountain plants show a total world distribution which cannot be classified into one of the three main groups. These distributions are very highly disjunct—that is, they consist of a number of areas

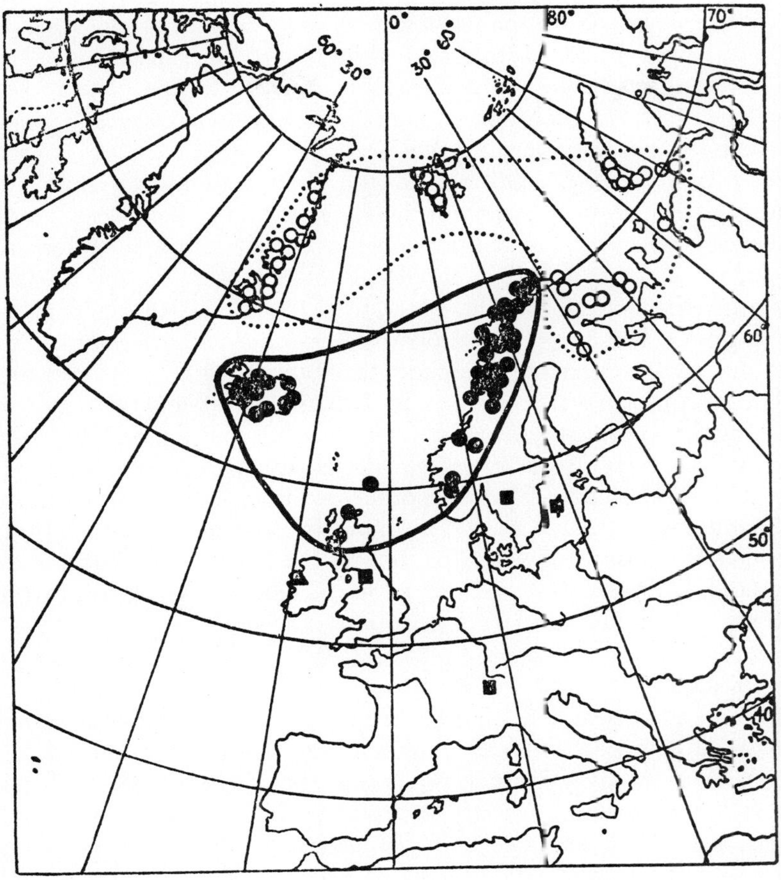

FIG. 8
World distribution of the mountain sandworts (*Arenaria ciliata* agg.)
(●) *A. norvegica* (▲) *A. ciliata* ssp. *hibernica*
(■) *A. gothica* (o) *A. ciliata* ssp. *pseudofrigida*
(after Melderis, 1953, in Lousley (ed.), *Changing Flora of Britain*, p. 91)

widely separated from each other and from the main area, if such exists.

The shrubby cinquefoil, *Potentilla fruticosa*, extends over a very considerable area of the Northern Hemisphere in all three continents;

but its European occurrences are very disjunct, ranging from the Pyrenees to the Baltic islands and absent from the Swiss Alps.

The rare grass *Agropyron donianum* occurs outside this country only in Iceland and N. America, though closely-related species occur in Scandinavia and elsewhere in the Arctic.

That most recent 'addition' to the British mountain flora, the Norwegian wormwood, *Artemisia norvegica*, possesses one of the most interesting of disjunct distributions, being known as yet only from the Norwegian mountains and in the Urals. Related arctic species seem to be quite distinct (cf. Hulten, 1954).

Finally a small group of sandworts, *Arenaria ciliata, A. norvegica* and *A. gothica*, possess striking disjunct distributions (fig. 8). The first-named, restricted to Ben Bulben in Ireland, has been distinguished from Central European plants of *A. ciliata* as a separate, and therefore endemic, subspecies, ssp. *hibernica; A. norvegica,* occurring in a few places from Shetland to Rhum, is otherwise known only from Iceland and Scandinavia; whilst *A. gothica*, found in Britain only round Ingleborough, occurs in the Swiss Alps, in Sweden and on the Baltic island of Gotland, from which it takes its name. These sandworts are further discussed in Chapter 8.

5. ENDEMICS.

We do not possess a single case of an indisputable endemic mountain species, sharply and easily demarcated from all other species. *Sagina boydii* is certainly distinct, but (see p. 124) there is a great mystery surrounding its origin. *Primula scotica* is hardly a mountain plant—though, as we shall see (Ch. 8), in Northern Scotland the distinction between mountain and lowland plants becomes virtually unworkable—and, in any case, taxonomically (though not cytologically) it is only with considerable difficulty separated from the Scandinavian *P. scandinavica.* There are undoubted endemics in the critical groups (e.g. *Euphrasia* and *Hieracium*), as we shall see later (pp. 74-6), but these are hardly comparable in origin with normal sexually-outbreeding species, and many would deny the category of species to these exceptional kinds.

This negative fact—the absence of clear endemism—is of interest and importance in understanding the origin of our mountain flora, which we shall be considering in the next chapter.

B. Perkins

Plate 5. Least willow, *Salix herbacea.* Ben Lawers

B. Perkins

Plate 6. Drooping saxifrage, *Saxifraga cernua.* Ben Lawers

FERNS AND FERN-ALLIES (PTERIDOPHYTA)

Although we are naturally primarily concerned, in a book entitled 'Mountain Flowers', with the flowering plants, we have thought it inadvisable to restrict our subject so as to exclude the other vascular plants. This is partly for traditional reasons (the ferns, club-mosses and horsetails have always been included in British Floras, and field botanists have studied them equally with the flowering plants), and partly because in the mountains these groups of more primitive plants do play a rather more obvious rôle than in most other habitats in the British Isles. In fact, to a botanist reared in lowland England, a visit to an upland or mountainous part of Britain may be more impressive for the richness and variety of the Pteridophytes than for any single group of flowering plants. This is not, of course, primarily an effect of altitude; it reflects principally the milder and more humid climate of the upland West and North. Nevertheless, the Pteridophyte flora of Britain contains quite a high proportion of species which can occur above 2,000 ft., and some few of considerable distinction as true mountain plants. The true mountain ferns we shall mention specially in the appropriate chapters in part II; they include the *Woodsia*, the mountain bladder-fern (*Cystopteris montana*), the parsley-fern (*Cryptogramma crispa*), the holly fern (*Polystichum lonchitis*, Pl. XIV, p. 149) and the alpine lady-fern (*Athyrium alpestre*).

Most British mountains provide damp, shaded habitats, often at quite high levels, by the sides of streams which have carved deep, narrow gorges, and in these a profusion of ferns may be found. Several of these, such as lady-fern (*Athyrium filix-femina*) and male fern (*Dryopteris filix-mas*), are common in many damp, shady, lowland habitats, but others, such as the delicate oak and beech ferns (*Thelypteris dryopteris* and *T. phegopteris*), are rare or absent from many lowland counties. The most wide-ranging species occurring in suitable habitats from the lowlands to near the tops of Highland mountains are the common bladder-fern (*Cystopteris fragilis*), the hard fern (*Blechnum spicant*) and the buckler-fern, *Dryopteris austriaca*. It has recently been shown (Walker, 1955) that small forms of the last-named species from Scottish mountains differ from the widespread British form in chromosome number; they are diploid whilst the common form is tetraploid. Similarly, Manton has recently (1950) shown that the common male fern *Dryopteris filix-mas* really consists of three more or less satisfactorily distinguishable species, one of which, *D. abbreviata*, is a diploid which appears to occur in this country typically if not exclusively on mountain scree slopes. It is interesting that, as with *Dryopteris austriaca*, the more local mountain form turns out to be diploid, whilst the form that is widespread chiefly in the lowlands is tetraploid. These cases afford interesting parallels to those in flowering plants (see p. 54) where races with differing chromosome number occupy mountain and lowland habitats.

The so-called 'sweet mountain fern' (*Thelypteris oreopteris*), although distinctly more common in the North and West of Britain, is perhaps not too happily named, for it has been recorded in all but a very few vice-counties in the British Isles, and has been seen recently in such lowland regions as Lincolnshire, East Anglia and Kent. Nevertheless, it is nowhere really common in lowland England, and a visit to a mountain or moorland area is recommended to see it growing really well.

The six British species of clubmoss (*Lycopodium* and *Selaginella*) are all, with one exception, northern mountain and moorland plants, and even the exception, the inconspicuous bog clubmoss *Lycopodium inundatum*, is a plant which I for one associate with the Scottish Highlands, for my first acquaintance with it was made by the shores of Loch Morlich under the slopes of the Cairngorm range. These remarkable plants represent primitive stocks whose ancestors flourished in the Carboniferous age long before any of the flowering plants are in evidence in the fossil record. The upright fir clubmoss (*Lycopodium selago*) is perhaps the most widely distributed of the group on British mountains, and occurs or occurred as far south as Cornwall, Devon and Dorset. It reproduces, as we have already mentioned, principally if not almost exclusively by means of bulbils. All the other British species of *Lycopodium* are creeping plants, whose main spread is obviously by simple vegetative means. The long stems of the stag's-horn clubmoss (*L. clavatum*) are familiar and decorative objects on many moorland stretches in upland Britain; the very success of the creeping, rooting habit makes the plant peculiarly easy to collect in long festoons, and it has undoubtedly suffered much from casual picking and should be admired and left to grow in peace. The interrupted club-moss (*Lycopodium annotinum*, Pl. 11, p. 124) is in this country much more strictly a mountain plant than the stag's-horn clubmoss which it resembles; it is a very rare plant in Northern England, but more abundant in parts of the Scottish Highlands (see p. 114). The alpine clubmoss (*L. alpinum*) is quite distinct, with its flattened fan-like branching creeping shoots, and is to be found at high levels in all the main mountain areas.

The clubmosses, like the ferns, have a complex life-cycle. The spores shed from the fertile 'cones' germinate to give inconspicuous structures known as prothalli, on which the sex organs are borne; after fertilisation of the egg-cell on the prothallus a young plant develops. Very little is known about the occurrence of prothalli in this country; indeed it would appear that they have only been seen on three occasions, and then those of two species only, *L. clavatum* and *L. selago* (see Manton, 1950, p. 245). This is no doubt partly due to their small size and peculiar nature—they are colourless structures only a few millimetres in length which develop more or less buried in the soil—but it seems probable that they are really quite rare and that virtually all young plants of *L. selago*, for instance, develop from the shed bulbils. A further problem is that, with the exception of *L. selago*, the bulbils of which will fairly easily germinate on *Sphagnum* moss, the British species seem to be very difficult to grow in cultivation. It is especially difficult to get them to produce ripe spores; and, of course, cultivation is necessary for experimental investigation.

In spite of these and other difficulties, Manton was able to determine the chromosome numbers of the British clubmosses, and showed that these justified the assumption that they were all probably only very distantly related to each other with the exception of the morphologically similar pair *L. clavatum* and *L. annotinum*. The most remarkable situation was found in *L. selago*, which has a very high chromosome number (c. 260) and shows marked irregularities at meiosis, suggesting strongly that this species has had a hybrid origin. If this is the case, the hybridisation must have taken place in the distant past, for there are no very obvious candidates as parental species, and certainly none living in the British Isles today.

The remaining British clubmoss, *Selaginella selaginoides*, is a diminutive plant which can very easily be overlooked or indeed mistaken for a moss. It often grows in the company of mosses by rivulets and in open marshy ground on mountains, and is particularly common where the parent rock has a reasonably high base content. It is a true arctic-alpine species, the only member of a large genus to show this type of distribution, and, with its short erect shoots bearing simple spirally-arranged leaves, more closely resembles *Lycopodium selago* in its general appearance than it does many tropical *Selaginella* species with flattened shoots. *Selaginella* differs from *Lycopodium* fundamentally, however, in bearing spores of two kinds; and Manton's work has shown that three different European species (one of them ours) have the same low chromosome number of $2n = 18$, which suggests that the species are naturally related together and reinforces the distinction from the *Lycopodium* species.

Finally, a word about the horsetails, *Equisetum*. None of the British species is exclusively a mountain plant, though several occur well above 2,000 ft. on British mountains, and one, *E. variegatum*, has a world distribution which is broadly arctic-alpine. Its British stations are rather scattered, in open, base-rich communities, chiefly in upland regions and in dune-slacks in the north and west; it shows, in fact, a distribution similar to the maritime group of mountain plants to be discussed later (p. 87), and its occurrence at sea-level may well be relict from the Late-glacial period. Dwarf forms of the common horsetail, *E. arvense*, occur not uncommonly on mountains, and some of these seem to differ so markedly from the typical form in stature and habit of growth that they have been distinguished as var. *alpestre*. Nothing is known, however, of the nature of their differences from the lowland weed forms. Two other horsetails, the delicately-branched *E. sylvaticum* and the rather similar *E. pratense*, are species with a northern distribution in Britain; the former is quite widespread, whilst the latter is curiously rare and far more selective in its habitats. Indeed, two of its best-known localities, Upper Teesdale and Glen Fee, Clova, are, as we shall see later, famous for their rich mountain flora.

With the recent confirmation of the occurrence of *E. ramosissimum* in Lincolnshire, only one European species remains undetected in the British Isles. This is the tiny arctic *E. scirpoides*, widespread in Northern Scandinavia and Iceland, and with a single not recently confirmed Alpine station. It resembles *E. variegatum* closely enough to make it a not unreasonable hope that a sharp-eyed botanist may yet find it on a Scottish mountain. We commend it as an exercise in botanical detection!

CHAPTER 4

ORIGIN OF THE MOUNTAIN FLORA (M.W.)

It is always interesting to trace back in the history of any subject the development of an idea which made an important contribution to understanding at an early stage; and it is surprising how often, when one attempts this, one finds that it is by no means clear to whom the full credit should be given, although a particular name is often associated with the idea concerned. The explanation of the striking similarity between arctic and mountain floras affords a very good illustration. The credit for this explanation is usually given to Edward Forbes, a remarkable scientist who made outstanding contributions to several branches of natural history, and compressed into a lamentably short life of 40 years the occupation of the chair of botany at King's College, London, that of natural history in Edinburgh, and the Presidency of the Geological Society. There is, however, an interesting passage in Charles Darwin's autobiography—a private document not written with any intention of publication (Darwin, 1888, vol. 1. p. 88)—where he is discussing the background to his great work 'The Origin of Species' (1859). It is worth quoting this in full:—'I was forestalled in only one important point, which my vanity has always made me regret, namely, the explanation by means of the Glacial period of the presence of the same species of plants and of some few animals on distant mountain summits and in the arctic regions. This view pleased me so much that I wrote it out in extenso, and I believe that it was read by Hooker some years before E. Forbes published his celebrated memoir on the subject. In the very few points in which we differed, I still think that I was in the right. I have never, of course, alluded in print to my having independently worked out this view.'

It throws a revealing sidelight on Darwin's character that he should have been content to refer generously to Forbes' work in the 'Origin' without mentioning that he had developed independently essentially the same view some years before Forbes made known his ideas.

Forbes first expounded his views in an historic contribution to the meeting of the British Association in 1845, and elaborated them in a paper published in the following year (Forbes, 1846) under the impressive title: 'On the Connexion between the Distribution of the existing Fauna and Flora of the British Isles, and the Geological Changes which have affected their area, especially during the epoch of the Northern Drift'. Here Forbes clearly formulated the principle of 'specific centres'—i.e. that every species must have had a particular centre of origin, and its present distribution is a result of dispersal from that centre—and, on the basis of this assumption, considered 'modes in which an isolated area may become peopled by animals and plants'. These he thought to be three, viz. 'by special creation within that area', 'by transport to it', or 'by migration before isolation'; and he gave cogent reasons why he rejected the first two and selected the last as being the method by which the British Isles have 'chiefly acquired their fauna and flora.'

Forbes saw what an important bearing the mountain flora of Britain, with its close similarities to the Arctic and the Alpine floras, had on these questions. In his conclusions he states: 'The alpine floras of Europe and Asia, so far as they are identical with the flora of the Arctic and sub-Arctic zones of the Old World, are fragments of a flora which was diffused from the North . . . ', and further: 'The termination of the Glacial epoch in Europe was marked by a recession of an Arctic fauna and flora northwards'. Forbes read the evidence with extraordinary insight; there is, as we shall see, nothing in our knowledge today which would lead us to disagree with, or even very seriously to modify, these two hypotheses.

Chapter eleven of the *Origin of Species* contains Darwin's detailed exposition of the significance of the remarkable facts of geographical distribution, and in particular of the resemblance of Arctic and mountain floras, as evidence for large-scale migration of floras and evolution of new types in isolation. To appreciate how well-formulated these ideas were in Darwin's mind at a much earlier date, however, it is well worth looking at two essays of his written in 1842 and 1844 (Darwin, C. ed. Darwin, F. 1909). The following passage from the later of these, for example, anticipates with remarkable vision the modern understanding of the spreading of the Late-glacial flora (see p. 84):—'I will only hazard one other observation, namely that during the change from an extremely cold climate to a more temperate one, the conditions, both on lowland and mountain, would be singularly favourable for the diffusion of any existing plants which could live on land just freed from the rigour of external winter; for it would possess no inhabitants; and we cannot doubt that *preoccupation* is the chief bar to the diffusion of plants'.

Another early contribution to this subject was also made by a famous English botanist, Sir J. D. Hooker, who read a paper before the Linnean Society of London in 1860 (Hooker, 1861). In this there is a detailed analysis of the Arctic flora in terms of the total world distribution of its component species; the extraordinarily wide occurrence of so many of the arctic species was fully appreciated, and Hooker considered also those most remarkable of distributions, the so-called 'bipolar' genera and species, which occur in the Southern Hemisphere as well as the North, separated by thousands of miles of temperate and tropical zones climatically unsuitable for them (see p. 60 and fig. 6).

Many of the main problems with which this chapter is concerned have already therefore been discussed for about a century. But, as we shall see, certain new lines of enquiry have only lately been opened up which throw a great deal of fresh light on the old questions.

Several types of evidence can be brought to bear upon the problems of the origin of the mountain flora of the British Isles. Only one of these is direct, namely, the elucidation of fossil and sub-fossil remains to throw light on the past distribution of modern mountain plants. The indirect evidence is provided by comparative studies of mountain and arctic plants, from several points of view, viz. taxonomic, cytogenetical, ecological and distributional; such evidence requires interpretation, a process which may often introduce a very considerable speculative factor. There is a good deal of literature bearing witness to the attraction of such speculation to botanists—and to the vigour with which a particular reading of the evidence has been upheld against others no more and no less speculative!

Direct evidence has, until recently, been almost non-existent, so that speculative interpretation of a variety of data has been able to go unchecked. We can conveniently proceed by looking at the types of indirect evidence, and then see how far modern sub-fossil studies in particular support, modify or refute conclusions tentatively reached.

The traditional arguments of plant geographers—arguments used by Forbes and Darwin—are based upon the assumption that plants of the same species occurring in widely separated areas must have had a single origin in some one area (not necessarily, of course, any of the areas now occupied by the species). This is not seriously questioned as a generalisation by present-day botanists, although it is realised that it is by no means necessarily true of some types of species (e.g. polyploid species). The presence of purple saxifrage in the Arctic, on a Welsh mountain and in the Swiss Alps can then only be explained in terms of dispersal or migration; either seeds (or other suitable parts) have been effectively transported from, say, the Arctic to the Welsh and Swiss mountains, or the saxifrage has in past ages spread gradually over the territory lying between the Arctic and the European mountains—not necessarily, of course, occupying all the intervening ground at any one period. The first explanation, in terms of efficient long-range dispersal, does not assume any sort of change in topography, climate, or otherwise throughout past ages; it is the simpler hypothesis and obviously cannot be disproved. The 'migration' type of hypothesis assumes a very considerable change in climatic and other factors in the earth's history, introducing the idea that many plants were able to grow in past ages, for climatic or other reasons, in areas which are no longer suitable for them. This dynamic concept of variation in plant distribution with variation in climatic and other enviromental factors seems in general to fit more of the facts concerning the past history of the earth, and we owe to Forbes its first formulation in general terms for the mountain floras.

Of course, the two hypotheses are not mutually exclusive; both gradual migration and long-range dispersal can be operating at the same time. We can, however, choose to assign the most important rôle in the establishment of a particular distribution to one or other; and, clearly, if we interpret discontinuous distributions as entirely due to chance successful long-range dispersal, we shall not assign to them any significance as indicators of the past history of the flora.

There has been a tendency, in discussions of distribution phenomena, to talk of an 'ecological' interpretation and a 'historical' interpretation as mutually exclusive. The 'ecological' view would point to the similarity of the habitat conditions between, say, the Arctic and the mountain habitats of the purple saxifrage, and imply that, given a particular set of environmental conditions, the particular plant will be found there. This view, of course, in its extreme form, implies complete efficiency of dispersal, and is quite untenable. The 'historical' view, on the other hand, stresses the importance of the past history of spread of the species in determining its present area, and, in its extreme form, implies that present-day ecological factors are of little or no importance in interpreting distribution; such a view is equally unsatisfactory. A balanced interpretation must include both historical and present-day ecological elements; if either is ignored it will be incomplete.

Most plant geographers have argued that 'accidental' long-range dispersal is insufficient to account for the abundant discontinuous distributions so characteristic of the European mountain flora, and have felt the necessity of interpreting them generally as 'relict' distributions of plants formally much more widespread. As we have already seen, the reproductive biology of many mountain plants would support the view that they represent relict populations which are managing to survive, often very precariously, on a small and specialised habitat which is all that remains suitable for them. The drooping saxifrage (*Saxifraga cernua*) has, in spite of the depredations of a century of botanising, just held its own in the same small area of mountain rock on Ben Lawers, reproducing entirely by bulbils; and no-one who has seen *Artemisia norvegica* on its exposed Scottish mountain ridge can possibly hold the view that this plant has recently arrived there by long-range dispersal by human or natural agency!

In the detailed interpretation of arctic-alpine distributions, however, the speculative element becomes so large as to make the exercise of doubtful value. In particular, one assumption often made seems highly dangerous; that is, that present-day distributions can give any simple clue to past migration of the species concerned. To take a concrete case, that of the shrubby cinquefoil (*Potentilla fruticosa*); to argue that its European localities represent a series of 'stepping-stones' by which the species reached its surprising western outpost on the limestone of County Clare is manifestly absurd, yet this type of argument has been constantly adduced in plant geography. It is even very questionable, in view of what we now know of the magnitude of post-glacial change involving extinctions of many species over wide areas, whether arguments from absence can be usefully employed—as, for example, the general argument that arctic species not now growing in the Alps never did grow there, or that species such as the Norwegian sandwort (*Arenaria norvegica*), which is now restricted to Scotland, Iceland and Scandinavia, have not been much more widespread at some relatively recent period. Indeed, as we shall see later, we are now able to show that some at least of these assumptions are quite untrue.

So far, we have assumed the origin of the mountain species in some distant age and presumably not where we find them on our mountains today. This seems to be justified in view of the virtual absence of endemics among British plants, which suggests that our flora is a recent, immigrant one, derived from the richer floras of the European Continent; and, as we shall see later, a good many arctic-alpine species have sub-fossil records of considerable antiquity. It is only in a few groups, where the distinctions between species have been for long a matter of dispute between taxonomists, that endemic species can be said to occur in the British mountain flora. The most interesting of these 'critical' groups in this connexion are the semi-parasitic yellow rattles (*Rhinanthus*) and eyebrights (*Euphrasia*), both members of the figwort family, *Scrophulariaceae*, and the hawkweeds (*Hieracium*) and dandelions (*Taraxacum*), which are familiar yellow composites.

In the case of the eyebrights, 24 species are described in the new British Flora, of which two in particular, *Euphrasia rivularis* and *E. cambrica*, are mountain plants thought to be endemic to this country (see p. 110). (In *Rhinanthus* the pattern of variation appears to be similar, but there is far greater uncertainty about the taxonomic situation). This local endemism in *Euphrasia* (and perhaps in *Rhinanthus*) is to be correlated with the occurrence of habitual self-pollination in the smaller-flowered northern or mountain species, in contrast to the cross-pollination of the larger-flowered, more robust lowland species (cf. p. 46). Theoretically, the effect of habitual self-pollination should be the production of local populations which are more or less 'pure lines', in which all the individuals are genetically very similar to if not identical with each other. (This situation contrasts with that of a normally cross-pollinated species, in which the individuals vary a great deal in genetic constitution, and the species may be composed of numbers of more or less separate populations between which, however, cross-pollination may be reasonably likely). A 'new' eyebright which arises locally, therefore, may perpetuate its kind exactly by self-pollination, and if this new type is particularly well suited to a rather special habitat, it may become locally common in a relatively short time. This theoretical possibility becomes more important in view of recent work on the eyebrights by Mr. P. F. Yeo (1956), who has produced evidence for the origin, by hybridisation between widespread species, of new types which could become locally established in just this way.

The other groups, the yellow composites, illustrate quite a different phenomenon. Here we are dealing with apomictic species, whose seed production does not normally involve any sexual process, and is therefore, biologically speaking, mere 'disguised' vegetative propagation. The seeds of an 'alpine' hawkweed or dandelion are therefore equivalent to, say, the bulbils of *Saxifraga cernua*; they reproduce the parent plant exactly, and the offspring have precisely the same genetic constitution as the parent. As with self-pollination, this situation results in a pattern of variation in which local populations are more or less distinct from all others, and more than 250 'micro-species' of *Hieracium* have been distinguished in this country alone, many of them endemic. We have far less detailed knowledge of the variation in dandelions in this country; no-one has seriously endeavoured to apply in Britain the taxonomy worked out for the group in Scandinavia, although Turrill (1938) has pointed out how and why this might be done with advantage.

Among the mountain flora, the hawkweeds of the groups *Alpina* and *Subalpina* have long excited a particular fascination for British botanists (see Ch. 1 pp. 22-3), and we have therefore accumulated a large body of information on the distribution of the many different micro-species. Sufficient information is now available to make it clear that, as was first pointed out by Scandinavian botanists, these apomictic groups are of very special interest for the study of plant geography in Northern Europe. Thus it seems clear that among *Hieracium* endemics there are essentially two different kinds; one, exemplified by the group of Shetland endemics each restricted to a particular stretch of cliff, which may well have arisen relatively recently more or less where they grow today, and another kind, exemplified by the discontinuously-distributed *H. lingulatum*, which are almost certainly 'relict' endemics formerly more widespread, whose origin we must assume to be very much older than the recent or 'neo-endemics'. The relict endemics presumably differ from other *Hieracia* which occur in scattered localities, say, in the British and Scandinavian mountains only in the degree to which their former areas have been reduced; and in theory, the end-point of such a reduction in range could be a restriction to a single cliff. These considerations suggest that endemic apomicts occurring in one locality only *may* be either recent or relict; and it is in the present state of knowledge pure speculation to assign particular cases to one type or another.

It is interesting to contrast the hawkweeds with another apomictic group well represented in the mountain flora, namely the lady's mantles, *Alchemilla*. In this genus the familiar British mountain plant is *A. alpina,* which is so far as we know totally apomictic throughout its wide range including the Alps and parts of the Arctic. Surprisingly, however, it is remarkably uniform throughout its North European and Arctic range, and the British plant is indistinguishable from, say, the Greenland and Scandinavian one; whereas in the Alps a whole group of closely-allied '*alpina*' forms occur, of which our northern plant is one. These facts can reasonably be interpreted only as indicating, firstly, the considerable age of *Alchemilla alpina* as a separate apomictic species, and, secondly, its recent invariability. The other *Alchemillas,* particularly the mountain ones, show the same sort of situation; thus *A. glomerulans* (Pl. XV, p. 156; fig. 9) has a total arctic-alpine distribution precisely parallel to that of a normal sexual species, and when seen on Scottish mountains is indistinguishable from Scandinavian plants, not only in general appearance and detailed characters, but also in choice of habitat and associated species. British *Alchemillas,* indeed, illustrate all three types of mountain plants (p. 58), for *A. conjuncta* is 'alpine', and *A. wichurae* is 'arctic' (though with an outlying European mountain station in the Riesengebirge).

The essential similarity of these distribution patterns to those of more familiar sexual species makes it necessary to produce the same hypothesis to account for both. In other words, we are bound to conclude that for thousands of years these apomictic *Alchemillas* have been distinct and essentially unchanged in form, though we shall see later that they must, in common with the whole mountain flora, have 'migrated' over many thousands of miles in Northern Europe.

In ordinary sexual species there is some indication of small-scale evolutionary change going on in the British mountain flora. A good many widespread species possess mountain races or ecotypes which could have differentiated more or less in situ, although in view of our knowledge of vegetational change we must feel that this is not necessarily the case. Such mountain ecotypes are commonly dwarfed in habit with relatively larger flowers than their lowland counterparts, differences which largely persist in culture and which are therefore based on genetic differences, which it is thought have gradually

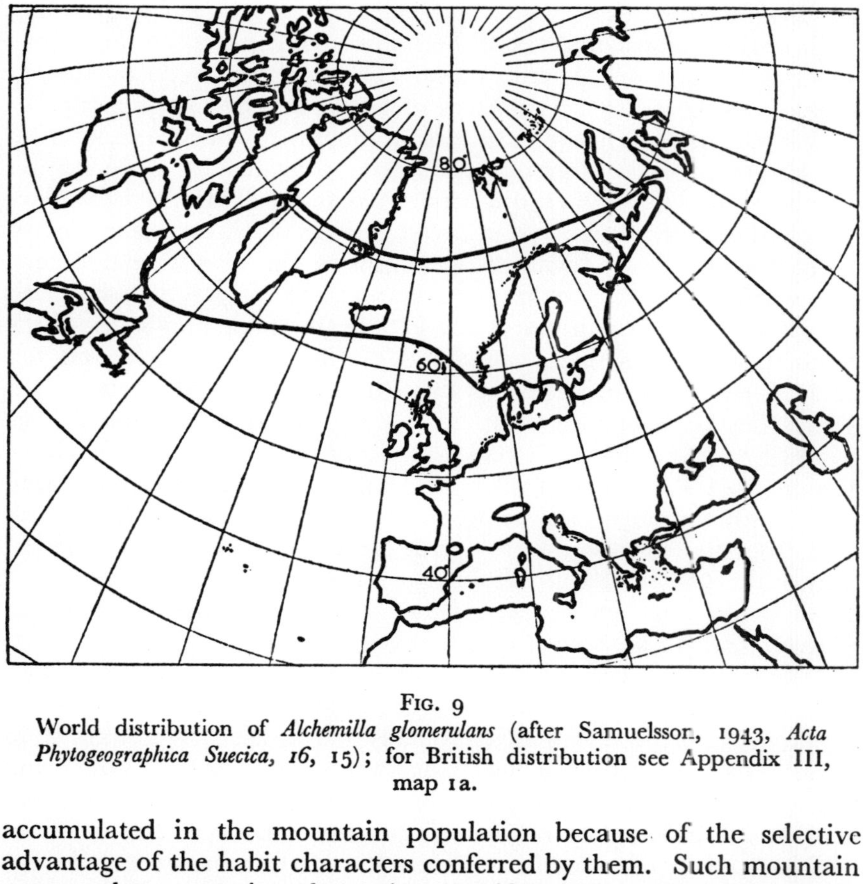

FIG. 9
World distribution of *Alchemilla glomerulans* (after Samuelsson, 1943, *Acta Phytogeographica Suecica, 16*, 15); for British distribution see Appendix III, map 1a.

accumulated in the mountain population because of the selective advantage of the habit characters conferred by them. Such mountain ecotypes have sometimes been given specific rank (e.g. *Solidago cambrica*), sometimes sub-specific (e.g. *Veronica serpyllifolia* ssp. *humifusa*), sometimes varietal or formal (e.g. *Leontodon autumnalis* var. *pratensis*, *Ranunculus acris* f. *pumilus*), depending more upon the taste of the taxonomist than the degree of difference from the type.

A small group of mountain plants of particular interest from the point of view of their history are those which have also a distribution around the sea-coast; the three best-known in Britain are the thrift or sea-pink (*Armeria maritima*), the sea plantain (*Plantago maritima*), and the sea campion (*Silene maritima*), which occurs up to at least 2,000 ft. in Scotland. Both the plantain and the campion have been the subject of detailed experimental taxonomic study, which has yielded a great deal of information about the nature of the variation they show. (cf. Gregor, 1938, and Turrill, 1948). These species will be considered later, when the post-glacial history of the mountain flora is discussed.

We are now in a position to turn to the direct evidence concerning the history of the British mountain flora provided by that rapidly-advancing study known as 'quaternary research'. This is not the place to describe in detail the type of investigations which are being carried on co-operatively by botanists, geologists, archaeologists and others, and which are so successfully elucidating the history of the British flora; a simple description of the general methods and some of the results can be found in Gilmour and Walters (pp. 48—52), and any botanist especially interested in this study could not do better than obtain Dr. Godwin's excellent new book, *History of the British Flora* (Godwin, 1956), which is a completely up-to-date and comprehensive account by the acknowledged British authority. Quaternary research has already provided us with an outline of the changes in the British flora from the time of the last great glaciation of Northern Europe. This outline, we can confidently expect, will be rapidly filled in in detail in the next few years, and many of the speculations of plant geographers will be proved or disproved.

The picture as we have it now is of an 'Arctic' flora at the edge of a great ice-sheet which covered much of what is now Britain for many thousands of years and which probably began the first halting steps towards its final retreat some 15,000 years ago. At that time, and indeed for several thousand years longer, the English Channel did not exist, and Britain was part of the Continent of Europe. At two main sites in Southern England, at Barnwell near Cambridge and in the Lea Valley, North of London, it has been possible to investigate deposits which undoubtedly date from this 'Full-glacial' period and must be more than 20,000 years old. Remains of plants—fragments of leaf, stem, flower, seed and fruit—have been painstakingly identified, and no fewer than 20 mountain plants still growing in the British Isles have been recognised at these two sites. These include *Dryas*, purple saxifrage, mountain cinquefoil, mossy cyphel, and several mountain sedges and willows. In addition to these, however, there are at least eight species of arctic or alpine distribution-type which are not to be found in the British Isles at all at the present time; these include the alpine poppy (*Papaver alpinum*), an alpine gentian, *Gentiana cruciata,* and the polar willow (*Salix polaris*). Clement Reid also described, from the Lea Valley beds, two species which he claimed were unlike any living plant; these were a flax (*Linum*) and a catchfly (*Silene*), which he named *Linum praecursor* and *Silene caelata.* In neither case, however, has re-examination of the material upheld Reid's view. The fossil flax seeds very closely resemble the living *Linum perenne* (Reid, E. M., 1949), whilst the material on which *Silene caelata* was based consists of a capsule and seeds indistinguishable from *Silene maritima,* associated in error with unrelated plant material thought to be calyces by Reid (Walters and Lambert, 1956). There seems, therefore, to be no clear case of a totally extinct species in these Full-glacial floras.

We have, then, indisputable evidence that over part of what is now Southern England there was, more than 20,000 years ago, a flora which bore striking resemblances to that occurring in the Arctic regions or high on the European Alps today. Thus the speculations of the early students of plant distribution are completely justified; the close similarity of Arctic and Alpine floras is now *known* to be due to their having been quite recently (in geological time) part of the same 'peri-glacial' flora which must have extended over wide areas of the European continent, and a great migration of vegetation must have taken place since the final retreat of the ice. In the course of this migration with climatic improvement (which we shall follow in more detail later), many species have become extinct over quite large areas, but have generally found odd 'refuges' on mountains where soil and climate conditions are most suitable for them; certain species have naturally failed to do so in the restricted opportunities afforded by the British mountains and these are now extinct here; and one or two may perhaps have become totally extinct, like the mammoth and the woolly rhinoceros which roamed the countryside in the Full-glacial period.

It would, however, be misleading to give the impression that the Full-glacial flora contained nothing but arctic or arctic-alpine plants. Plants familiar in lowland Britain today appear in the lists, particularly species growing in marshy ground or water, such as the greater spearwort (*Ranunculus lingua*), the marsh cinquefoil (*Potentilla palustris*) and species of pondweed (*Potamogeton*), and also plants of disturbed ground (present-day weeds or roadside and waste-ground species) such as chickweed (*Stellaria media*) and knotgrass (*Polygonum aviculare*).

Several types of plant are absent from the Full-glacial floras. Trees and shrubs are obvious cases; the only species represented is juniper, and the dwarf mountain form of this plant (which was presumably the form present in this Full-glacial flora) is a very low-growing shrub indeed. This is, of course, the situation we find in the treeless Arctic tundra today. Another category of absentees, which is to be expected, comprises the present-day woodland plants—for example, the familiar dog's mercury (*Mercurialis*) or bluebell (*Endymion*). A third category is represented by the heaths (cross-leaved heath, *Erica tetralix*, and heather, *Calluna*); these are plants with a present-day 'Atlantic' type of distribution, for which presumably the Full-glacial climate was too extreme (but see below).

The final stages in the retreat of the ice are represented by the so-called 'Zones I, II and III' of the pollen analysts' classification, which collectively constitute the Late-glacial period, occurring approximately between ten and fifteen thousand years ago. In Zone I, the 'older *Dryas* period', we have an arctic tundra flora, not essentially different from the Full-glacial flora; but this is followed by the Allerød period (Zone II), so-called from the site in Denmark where it was first recognised in detail, in which there was obviously a more favourable climate, suggesting a more luxuriant vegetation. Zone III, the 'younger *Dryas* period', represents a temporary re-advance of glacial conditions, with the restitution of an arctic tundra flora over much of North-West Europe. A good many deposits of Late-glacial age have by now been investigated, and more than 200 species of flowering plant have been identified from Zones I to III in the British Isles alone, so that we have a useful picture of the flora which followed the retreating ice in North-West Europe.

In Britain, as in Denmark, the vegetation, at least at the height of the Allerød period, can perhaps best be described as 'park-tundra'—a term due to the Danish investigator Iversen. We can picture a country recently scraped by ice and plastered with glacial debris, over which an open vegetation is colonising with scattered clumps of the pioneer trees and shrubs, the birch and shrubby willows. The main

H. E. Lewis (British North Greenland Expedition)

Plate 7. Moss campion, *Silene acaulis*, with arctic fritillary. North Greenland

M. C. F. Proctor

Plate 8. Alpine butterwort, *Pinguicula alpina*. Austrian Alps vegetation, to judge by the proportion of the different types of pollen in the Late-glacial deposits, was composed of grasses and sedges; but there is in addition a rich assortment of other herbaceous plants. In North-West Ireland especially, the crowberry (*Empetrum*) is very abundant in Zone II, whilst the least willow (*Salix herbacea*) is very characteristic of Zones I and III.

It is particularly in the identification of the constituents of this rich Late-glacial flora that rapid progress is being made. Already it is clear that this period was the heyday of many herbaceous species which, in the more recent history of the British flora, have been largely or in some cases entirely squeezed out of existence by the severe competition of the pine forest, and later the deciduous forests, which spread over the face of the land as the climate further improved. Genera such as *Artemisia* (mugwort, wormwood), *Campanula* (harebell, bell-flower), *Helianthemum* (rock-rose), *Rumex* (dock, sorrel), and *Thalictrum* (meadow-rue) all occur with high constancy and frequency in the pollen-analyses of the British Late-glacial floras. In these and similar cases, it is not possible to say which *species* were actually present, though we can recognise the genus with certainty from its pollen grains. A good many other identifications can be made more precisely, however; for example meadow-sweet (*Filipendula ulmaria*), devil's bit scabious (*Succisa pratensis*) and valerian (*Valeriana officinalis* agg.) are all common and characteristic Late-glacial plants. As with the Full-glacial floras, there is a high proportion of marsh and water plants on the one hand, and, on the other, of plants we think of as ruderals and weeds today. A countryside over which the retreating ice had left great lakes, large areas of flooded ground and piles of morainic debris would naturally offer to both these classes of plant an unparalleled opportunity for spread and establishment; and such a lush vegetation would offer in its turn grazing-grounds of exceptional richness to the large herbivorous mammals such as the elk, bison and the now extinct giant Irish deer.

So much for the general picture of the Late-glacial vegetation. Let us look more closely now at the evidence concerning the arctic and alpine elements in this flora. *Dryas*, as we have already seen from the names given to Zones I and III, is an abundant Late-glacial plant. Its abundance emphasises another characteristic of the flora of that time —namely that it contained many species (like *Dryas* itself) which we know today are more or less confined to soils which are not acid in reaction—so-called 'calcicole' species. This is understandable when we think of the great areas of chalky boulder-clay deposited by glaciation over the countryside, and also when we realise that 'young', newly exposed soils which have not been long subject to the leaching effect of rainfall naturally contain a higher proportion of the soluble bases which cause an alkaline reaction. This gives us the necessary clue to the understanding of the phenomenon already commented on—that on British mountains the really rich flora is always associated with the presence of basic rock, and that hard, acidic rocks and acid peat deposits are obviously inimical to the majority of our mountain plants, although in fact by far the greatest area of the British mountains affords only acid rocky and peaty soils. For the calcicolous members of our mountain flora, then, the Late-glacial period was one offering exceptionally favourable conditions; in addition to *Dryas*, one could name in this category the dwarf willow *Salix reticulata*, the hoary whitlow-grass (*Draba incana*), and the shrubby cinquefoil (*Potentilla fruticosa*) as characteristic Late-glacial plants recorded with certainty from more than one British site well south of their present restricted range.

The shrubby cinquefoil is a particularly interesting case, for, as we have seen earlier (pp. 63-4), it has one of the most remarkable disjunct distributions to be found anywhere in the world today. The knowledge that this plant was growing in the South of England some 10—15,000 years ago makes certain the 'relict' interpretation of its outlying localities, some of which, such as those on the bare coastal limestone of Western Ireland and the limestone 'alvar' of the Baltic Island of Öland, are lowland, whilst others, such as the Northern English Teesdale and Lakeland and the Pyrenean localities, are more truly mountain ones. Here is clearly a plant, formerly widespread and abundant, which has retreated before its more successful competitors—the shade-producing forest trees in particular—until its total European range is represented by a few small widely-separated areas where highly peculiar conditions have allowed its survival. It is certainly no accident that the Burren region of Co. Clare harbours such a peculiar flora, in which both *Dryas* and *Potentilla fruticosa* are locally abundant; this curious largely treeless vegetation of the Irish limestone must be a direct descendant of the Late-glacial vegetation of ten thousand years ago. It is an odd corner of North-West Europe over which the forest has never completely closed nor acid blanket bog developed;

and on it is collected together as remarkable an assortment of flowering plants as one could find anywhere in the world—the arctic-alpine *Dryas*, the northern continental *Potentilla fruticosa*, the Mediterranean orchid *Neotinea intacta*, and several others growing in close proximity.

If *Dryas* was the abundant calcicolous species of the Late-glacial flora, then the crowberry, *Empetrum*, was certainly its counterpart on the more acid soils. Crowberry occurs in Late-glacial deposits in England, Scotland and Ireland, but is particularly abundant in Irish Allerød sites. Understandably, it has not been forced to retreat so completely as *Dryas* in the modern British vegetation, for large areas of acid moorland still afford suitable habitats for it. (There are strictly speaking two crowberries (see p. 54); and it is interesting and perhaps rather surprising to find that the Late-glacial records so far seem to be referable to the diploid *E. nigrum* only). Unlike *Dryas*, the crowberry is absent from the Full-glacial flora. Similarly the heather and the cross-leaved heath appear for the first time in the Late-glacial floras; but this is more understandable, since they are both, and particularly the latter, more strictly 'Atlantic' plants whose spread northwards might well have been retarded by unsuitable climatic conditions near the ice-sheet.

Another characteristic Late-glacial plant is the dwarf birch (*Betula nana*), which, together with its taller relative the silver birch *B. pubescens*, contributes the abundant birch pollen to the Late-glacial deposits. The dwarf birch is today restricted in Britain to Northern Scotland (see p. 155) and is totally absent from Ireland; yet it must have been abundant at one time over large areas of the country. We have therefore evidence that in the last 15,000 years this species has become totally extinct in Ireland; and we now know this is also true for the sandwort *Minuartia stricta*, and for the alpine milk-vetch (*Astragalus alpinus*), which are Irish Late-glacial plants.

Until very recently no certain case was known of a Late-glacial plant now completely absent from the British Isles; but in 1954 Miss A. Conolly discovered in Late-glacial deposits seeds of a species of *Papaver* which can only belong to one of the arctic or alpine poppies, none of which now occur in this country; and it seems certain that it is only a matter of time before other cases are brought to light.

The weight of evidence points to extinction on a very considerable scale as a factor in explaining how relatively poor in species our mountain flora is compared with that of the Alps or even of Scandinavia, and also, within the British Isles, how much poorer the Irish mountain flora is than the Scottish. Several mountain plants not among the rarest class in Scotland are extremely rare in Ireland (see Ch. 10, pp. 184-5) for example, the alpine lady's mantle (*Alchemilla alpina*), and the alpine saxifrage (*S. nivalis*); and we can presume that many more arctic and alpine plants must, like the dwarf birch and the sandwort, have succumbed completely in Ireland to changing climate, soil and vegetation in the last 10,000 years.

We have already seen how remarkably frequent 'safe' self-pollinating vegetative or apomictic methods of reproduction are in British mountain plants, and further how the proportion of polyploids in the mountain flora seems to be significantly high. The conditions of Late-glacial times may again provide a key to the understanding of these facts. In effectively colonising ground progressively laid bare by the retreating ice a great advantage would accrue to the first arrivals, who could get well-established quickly. An unsafe method of pollination, dependent upon insect visitors, would be a real disadvantage. In the same way, familiar arable weeds today, such as the shepherd's purse (*Capsella bursa-pastoris*) and the chickweed (*Stellaria media*), are almost exclusively self-pollinated and thereby set abundant seed; others, such as the ground elder (*Aegopodium podagraria*), succeed by having a very effective means of vegetative spread, and yet others, such as the dandelion (*Taraxacum*), set seed apomictically. The resemblance, in terms of species, between these Full- and Late-glacial floras on the one hand and our modern weed flora on the other we have already commented on; and this again can be no accident. In many cases, the species which succeeded in the Late-glacial flora are those which survive today largely because man's activities provide them constantly with temporary 'open' habitats into which they can quickly and effectively spread.

The Late-glacial flora then might be characterised as a flora of 'opportunists', moving forward into the newly-available ground in the wake of the retreating ice. Factors of competition between one plant and another would be at a minimum. This stage, however, could not last long in any one place, no more than a neglected garden for long remains inhabited only by the arable weeds. A closed community would develop, first a *Dryas* or crowberry 'heath', and, later, scrub or birch woodland, and many of the smaller arctic-alpines would be either crowded out altogether or at least seriously reduced in numbers.

If 'safe' reproduction is such an advantage to plants colonising open ground, we might ask why the majority of flowering plants possess more or less effective modifications to ensure cross-pollination, with all the attendant risks? This is a subtle and complex question, to which, the majority of botanists would agree, we can as yet give only a partially satisfactory answer. To Charles Darwin and other nineteenth century writers the significance of sexual reproduction, involving cross-pollination in the flowering plants, lay in some mysterious 'rejuvenation' of the stock thereby achieved. This was not, however, a concept which could be put into scientific terms, and it was inevitable that it should remain vague until the growth of modern cytogenetic studies. Avoiding technical detail the modern view might be expressed as follows.

Firstly, there is no evidence in the plant kingdom that species reproducing exclusively by non-sexual means eventually 'die of old age'. Indeed the whole question of the length of life of a vegetatively-reproducing plant is mysterious, and all that can be safely said is that many plants relying exclusively on apomictic or vegetative methods of reproduction must be at least several thousand years 'old' and probably several hundreds of thousands. Thus, if *Saxifraga cernua* is indeed on the verge of extinction on Ben Lawers, we must look for some other, and partly external, explanation of its impending demise (cf. Holden (1952), p. 19, for the contrary opinion !).

Secondly, there is a fundamental difference between sexual reproduction involving cross-fertilisation and non-sexual methods. This is inherent in the process of formation (in flowering plants) of pollen and embryo-sac by what is known as meiosis. This process permits the recombination of the hereditary material contributed to the particular plant from its two parents; if this hereditary material is different, then new combinations can be produced, and the next generation will contain individuals among which a large proportion of the possible combinations of the available characteristics will be represented.

Thirdly, this potential variability of the sexual species would theoretically confer an advantage by enabling the species to adapt itself relatively easily to a *changing* environment. Plants possessing the particular combination of characters fitting them to the new conditions (whatever they may be) may occur by random recombination and may then succeed, thus perpetuating the species. The power of genetic adaptation of this sort has been lost by the apomictic or vegetatively-reproducing species. It looks, then, as if, in evolution, the short-term advantages of 'safe' adaptation to particular conditions conflict with the long-term necessity of retaining sexual crossing to ensure adaptability and the eventual survival of the species.

If this is true—and it should be understood that, however plausible they may be, such theories are very difficult to *prove*—then its relevance to the understanding of the British mountain flora is obvious enough. This remarkable collection of curiously-adapted plants which are scattered in odd corners of our mountains represents the remaining traces of an earlier, transitory flora which, by adapting itself closely to extreme conditions, had sacrificed the possibility of long-term survival. There *is* then a certain measure of truth in the statement that *Saxifraga cernua* is doomed to extinction because of its abandonment of sexual reproduction; for some slight environmental change, which a variable sexual species might 'take in its stride', could easily prove too much for the last constant unadaptable remnants of the saxifrage in the British mountains.

There is yet another aspect of the Late-glacial vegetation conditions which is of significance in understanding the present-day mountain flora. This is the exceptional opportunity which must have been afforded at that time for hybridisation between species which, before the great ice retreat, might have been evolving in isolation. Not only would ordinary species-hybrids with varying degrees of sterility be readily produced in the immense intermingling of different floras, but also by chromosome doubling new allopolyploid hybrid species would undoubtedly be found. A considerable proportion of common North European species we now know or can reasonably infer to be of allopolyploid origin, among them some of our most familiar and variable plants, such as the annual meadow-grass (*Poa annua*), the bird's-foot trefoil (*Lotus corniculatus*) and the common British wild thyme (*Thymus drucei*). It is probably significant that all three of these plants occur quite high on British mountains today; they are all likely constitutents of a Late-glacial flora, and could indeed have originated in this or a similar earlier period of immense vegetational 'upheaval'. As we have seen, polyploids are significantly high in proportion in the mountain and Arctic floras, and hybridisation seems to have been involved in the origin of important polyploid groups such as the mountain meadow-grasses (*Poa*) and the apomictic lady's mantles (*Alchemilla*).

Post-glacial Vegetation Change

The post-glacial history of British vegetation must have been, so far as the arctic-alpine Full- and Late-glacial floras are concerned, one of wholesale retreat and extinction. Over the main areas of Britain the birch forest of Zones IV and V was succeeded first by the Boreal pine forest, and then, after a climatic change to wetter conditions probably contemporary with the formation of the English Channel, by the broad-leaved forests of the Atlantic period. The remnants of the Late-glacial flora retreated to the treeless mountains and to scattered and diverse habitats which for various reasons remained free from closed woodland or scrub cover. In a good many of these habitats we can still find them today.

One of the most extensive is the sea-coast habitat, or rather series of habitats, ranging from salt-marsh to sand-dune, dune slack and sea-cliff. As the Scots naturalist Buchanan White appreciated over seventy years ago (White, 1879), this is undoubtedly the explanation of the remarkable similarity between the mountain and the maritime flora of Britain; we know, for example, that the sea-pink or thrift (*Armeria*) and the sea-plantain (*Plantago maritima*) were abundant Late-glacial plants, and might reasonably infer that the sea campion (*Silene maritima*) was likewise widespread in this flora—although in fact the only early records for the seeds of this plant are as yet Full-glacial ones. Indeed we can still point to *inland* habitats in the lowlands today where some of these plants must have survived from the Late-glacial flora as they have on mountains or on the sea-coast. One of the most interesting is on bare shingly ground on Ross Island in the Killarney lakes in Western Ireland, where thrift and sea-campion grow together on a very restricted area free from the shade of woody plants or serious competition from tall growing herbs. The distinction between the mountain and maritime flora is indeed very difficult to make in many parts of the north and west of the British Isles (see especially p. 155).

Other 'relict' habitats of this type, on which Late-glacial species of the mountain flora may survive, include limestone cliffs (e.g.

mossy saxifrage at Cheddar, or purple saxifrage on Ingleborough), inland marshes, particularly over limestone (e.g. *Bartsia alpina* in the Malham and Orton areas), and the sides of rivers (e.g. the shrubby cinquefoil and many other species in Upper Teesdale).

Although the closed forest has been the most obvious 'enemy' of the Late-glacial flora, it is by no means the only one. Another which in the more recent history of British vegetation may well have been responsible for a further decimation of the mountain flora, particularly in the West, is the development of acid peat deposits which, in conditions of high rainfall, can cover even limestone with blanket bog. The relict habitats we have pointed out are in fact all places where presumably neither forest nor bog could have developed at any period in post-glacial history. In this connexion it is important to notice that the vicissitudes of climatic change in the last 10,000 years have meant that there have been (different) periods when both the forest and the blanket bog extended more widely in many parts of upland Britain than they do today.

This may mean that for certain mountain plants the present-day situation has allowed a limited expansion from an extremely reduced state. Such 'bottlenecks', when for perhaps hundreds of years the population of a plant is reduced to very small numbers of individuals, are thought to have serious effects on the genetic constitution of the species; in such conditions, it can be shown, there is a danger that the genetic variability of the species may be so greatly reduced that, even though it is reproducing sexually, when the opportunity for expansion comes, it may be unable fully to take advantage of it. Some such explanation may underlie the extreme restrictions of certain British mountain plants, which are difficult to account for in terms simply of the conditions of their environment; though we must frankly admit our ignorance on all these matters, and treat such ideas as mere guesses until we have experimental evidence.

To complete the story of the mountain flora to the present day, we might ask whether there is any evidence that very recent climatic changes are having a measurable effect. There is undoubted evidence for a general small 'improvement' in climate over most of Europe in the past 100 or 150 years. Perhaps the most impressive indication of this is given by the quite spectacular retreat of glaciers and ice-fields in the Alps, Scandinavia and Iceland; and that there has been a comparable change in Britain is fairly certain, although the observations on Scottish snow-patches, for example, are less satisfactorily documented (see Manley, 1952). Further, there is very good evidence that some southern species of flowering plants have extended their British range recently; the lizard orchid (*Himantoglossum hircinum*) is the classic example (see Good, 1928). The difficulty in considering a corresponding retreat of northern species—which can in fact be demonstrated in a good many cases in lowland England—is to distinguish between decline and extinction which is due to human interference, direct or indirect, and that which is presumably due to subtle changes for which slight climatic improvement is responsible. Thus, the Alpine sedge *Carex davalliana*, formerly known at one locality near Bath, is now extinct; but we know that this was due to human activity—in fact the site was drained and built on about 1840! Similarly the alpine cotton grass (*Trichophorum alpinum*) used to grow in one bog in Scotland, but disappeared when the bog was drained, and the beautiful white-flowered alpine butterwort (*Pinguicula alpina*) (Pl. 8, p. 81) has apparently long since disappeared from its only known station in a bog in the Black Isle of Cromarty. The butterwort is, incidentally, a plant which may well be re-discovered elsewhere in the Highlands; it is difficult to recognise from the common species when not in flower, and may well flower quite early in the season.

Even more difficult to analyse are the cases of northern moorland species with isolated stations in the South of England, for which the records in local floras indicate a decline and gradual extinction; examples are the crowberry in Dorset, and the clubmosses *Lycopodium clavatum* and *selago* in Cornwall, Devon and Somerset. In such cases indirect interference factors of moorland management, such as grazing and burning, or even a mere change in the intensity of one of these factors, may be responsible for the decline of a species already narrowly restricted in its tolerance of the habitat conditions. The possibility must also be borne in mind that some at least of these isolated records of northern species south of their main British range may represent relatively recent arrivals rather than populations relict from the Late-glacial or early Post-glacial period. This seems particularly plausible as an explanation for the occurrence of the Scottish pine-forest orchid *Goodyera repens* in coniferous plantations in Norfolk, several hundred miles south of the nearest Scottish locality. Even here, however, the evidence is by no means conclusive, for native pine has undoubtedly persisted in East Anglia, if only in small quantity, throughout the Post-glacial and the recent historical periods, and we cannot be sure that the orchid is not a native stock which has a much longer history in Norfolk than the nineteenth century plantations.

Another northern orchid, the lesser twayblade (*Listera cordata*), shows a relict mountain and moorland distribution in England and Wales. It is perhaps originally a northern birch and pine forest herb—it certainly occurs in such communities commonly in Northern Europe—but it is mostly found in moorland and blanket bog communities extending to well above the present tree limit on our mountains. It is a very inconspicious plant which may well be commoner than the published records suggest; the pair of simple leaves and the small terminal spike of greenish flowers are quite unmistakable, but may never be seen at all from above until the heather, crowberry and bilberry layer is parted and the moss layer underneath studied. Thus its published records for the Berwyn range—which apart from an isolated patch still surviving on Dartmoor constitutes its known southern limit in Britain—are very scanty indeed, but field work this year (1955) by a party of Cambridge students studying the composition of the blanket bog on Moel Sych, the highest of the Berwyn group, revealed that the orchid was locally quite frequent there above about 2,200 ft.

The vast majority of these isolated occurrences of northern plants in lowland England have probably already been discovered, but not apparently all, for the year 1955 has produced two very interesting additions. The northern marsh orchid, *Orchis purpurella,* has been found growing near Southampton, and even more surprisingly the chickweed wintergreen, *Trientalis europaea,* has been recorded in Suffolk. *Trientalis* is a common Scottish species whose natural habitat, like *Listera cordata,* is the northern coniferous or birch woodland, but which will grow quite well on mountain moorland; its previous known southern limit in this country was Lancashire and Yorkshire. Both these occurrences might represent recent introductions; but the fact that they have never previously been recorded in lowland England is certainly not to be taken as evidence of their recent arrival, but rather as evidence of the incomplete state of our knowledge even of the flora of accessible lowland counties. It is particularly easy for the competent field-botanist, who knows roughly what species to expect to see in any particular region or habitat, to have a 'blind spot' for the unexpected! Both these recent discoveries, and several others with an equal or greater element of the unexpected, were made by members of the

Botanical Society making field lists for the Society's Distribution Maps Scheme (see Conclusion, p. 195). We can almost certainly expect a similar number of discoveries in future seasons, which may well somewhat alter our views as to the southern limit of other northern species.

Praeger (1950, p. 66) also gave evidence that the cloudberry (*Rubus chamaemorus*) has greatly decreased in its single Irish station in recent years and is now apparently extinct, and other northern species are on the verge of extinction in Ireland, presumably for climatic reasons. It is here that one most regrets the lack of satisfactory observations over a period of years on the variation in population size of *any* of our rarest mountain plants, for such a series of data would be of the utmost interest in this connexion.

Per-glacial Survival and the Glacial Periods

So far we have traced the origin of the mountain flora to the widespread Late-glacial floras of 10—15,000 years ago. It is obvious, however, that we cannot look to that period for the origin of the actual species of plants concerned except in a very few possible cases (allopolyploids and certain apomictic microspecies). The vast majority of arctic and alpine species—and indeed the majority of apomictic microspecies also—must be considerably older than this. The Full-glacial provided us with a rich assortment of modern arctic and alpine plants, and we must remember that the so-called 'ice-age' actually consisted in Northern Europe of several (possibly four) distinct glaciations, of which we have only so far mentioned the most recent. Between each of these glaciations there were interglacial periods lasting thousands of years, during which a temperate climate was again restored over North-West Europe. Active work is proceeding on the study of the flora of these interglacial periods, and sufficient information is already available about the character of the last inter-glacial to say that it showed a full climatic and vegetational cycle, beginning with arctic tundra, passing through coniferous to broad-leaved forest, and then back again to tundra. Clearly, if this process has been repeated, with variations, no fewer than three times in the recent history of the flowering plants of the Northern Hemisphere, we would expect our efforts to elucidate this history of change in terms of present-day plant distribution to be beset with great difficulties.

It is certain that the magnitude of these changes has been seriously under-estimated by those who have advanced theories concerning the diverse origin of arctic, northern and alpine floras. One of the boldest attempts of this sort has been made by the Swedish botanist Hulten (1937), who has an unrivalled knowledge of the Arctic flora. Arguing on the basis of present-day distributions, he has suggested that the main bulk of arctic and arctic-alpine species are very old, and that they had mostly established a circumpolar distribution before the onset of the ice ages. He pictures the effect of glaciation—and particularly that of the maximum glaciation—to have been to wipe out this flora over very large areas, leaving certain species surviving only on refuges north of the ice, and others surviving both here and to the south. The former group have developed into our present arctic species, the others into the arctic-alpine ones. He further asserts that the distributional evidence points to the Eastern Asiatic and North-Western American areas as being the ones from which the main arctic and arctic-alpine flora spread, and presumably where the main ancestral types originated in the Tertiary period millions of years ago.

The serious weakness in Hulten's arguments seems to be in his stress on the importance of the glaciations as such, as being responsible for large-scale 'migration' and extinction of species, to the neglect of the recent post-glacial migration and extinction phenomena. If we are to understand these complex patterns at all, we must, as it were, start from the present day and work backwards, on the reasonable assumption that the most recent history of the flora left the most obvious impression on it.

Thus, if we can explain the extraordinary disjunct distribution of the shrubby cinquefoil—a distribution whose interpretation Hulten confesses is not clear to him—in terms of widespread colonisation in the Late-glacial and subsequent extinction, there is no need to invoke any theories of 'per-glacial survival' to account for it. This does not mean that we should not be concerned with the earlier history of this plant, which was, in fact, present in the Full-glacial flora of Barnwell and is presumably of very ancient pre-glacial origin; but merely that we should proceed with caution from what we know of recent change into the extreme uncertainties of earlier vegetational history only as our evidence justifies or requires it. In the same way, the majority of cases of distributions, chiefly of mountain plants, which have been cited in the British literature as affording evidence of 'per-glacial survival' (cf. Salisbury and Wilmott, 1935), can be explained much more simply and satisfactorily in terms of Late-glacial spread and Post-glacial extinction.

A useful short account of the origin and development of the Scandinavian per-glacial survival theories and their application to the problems of interpreting the peculiar distribution of arctic and arctic-alpine plants in Britain was given by Melderis at the Botanical Society's Conference in 1952 (Melderis, 1953). This paper, with its very complete bibliography, would be most useful to anyone wishing to follow the discussions and arguments of botanists on this complex subject. As Godwin (1949) has shown, however, our recently acquired and rapidly expanding knowledge of the Late-glacial flora of Britain has revealed that much of this argument has been irrelevant as applied to the British flora. We now *know* that unglaciated regions of Southern England carried, during and after the last glaciation, a 'tundra' flora, containing many species of our present-day mountain vegetation; the question is no longer whether native British mountain species survived the last ice-age 'on British soil'—for Britain was part of the European continent anyway—but whether there is any evidence for supposing that certain species survived south of the ice-sheet while others survived on relatively restricted areas within the main glaciated zone. In both cases we assume a spread after the ice retreated. The evidence that particular areas in the British Isles which are thought on geological grounds to have escaped glaciation have peculiar rare or endemic plants is not very impressive; and few botanists would argue strongly today in favour of the idea, until recently rather widely held, that the peculiar collections of rare species to be found, for example, in the Upper Teesdale area have anything to do with the occurrence of 'nunataks' (see below) in the vicinity. Nor is the evidence concerning the flora of the Hebrides and other outlying islands particularly conclusive, though the case for local per-glacial survival in this region would seem to be somewhat more easily made (cf. Heslop-Harrison, J., 1947). Reduced to reasonable terms, therefore, the 'per-glacial survival' hypothesis, so far as it applies to the British flora, would appear to be presentable in some such way as this.

The majority of present-day British species, whether mountain or lowland, are obviously part of a re-immigrant flora from the European continent, a flora which appeared in successive 'waves' as the climate improved. Any species which can be shown to have been present in any stage of re-immigration (Late-glacial or Post-glacial) can be assumed to owe its present occurrence here at least in part to that re-immigration. A certain proportion of the flora, mostly but not exclusively of arctic or arctic-alpine type, may, however, have survived the last ice-age north of the southern extension of the ice-sheet on unglaciated land of greater or less extent, particularly along the western coast of the British Isles. Such unglaciated areas, known as 'nunataks', occur at the present time, for example, in those parts of Greenland largely covered by a permanent ice-sheet, and they can support a limited flora even under these extreme conditions. The British flora, and particularly the mountain flora, could then have been recruited from two sources after the last ice-age, namely the restricted 'per-glacial refuges' and the 'peri-glacial flora' of the main part of Europe just south of the ice-sheet.

Several kinds of distribution would seem to offer a basis for such a modified theory of per-glacial survival in the British flora. The most obvious kind would be species which have isolated occurrences on mountains in the north and west of the British Isles, and perhaps also in Iceland, Scandinavia and elsewhere in the Arctic, but which are otherwise completely absent from Europe; for example *Arenaria norvegica* (see fig. 8, p. 63) and *Koenigia islandica* (see p. 135). The argument would then run that the absence of these species from any part of the European mountain chain must be taken as evidence that they, unlike the arctic-alpines such as *Dryas*, were not present or did not survive long in the peri-glacial flora south of the ice-sheet, and therefore did not re-immigrate in the Late-glacial. It is certainly interesting in this connexion that the arctic willows, *Salix lapponum* and *S. arbuscula*, which have been identified in British Full-glacial deposits, have *not* so far been found in the Late-glacial.

Dahl (1953) has collected a good deal of evidence to support his view that many Scandinavian mountain plants are restricted in their present-day distribution by climatic factors, particularly the maximum summer temperatures. He further suggests that the absence of such plants from Late-glacial and Post-glacial deposits in Central Sweden, though they are present in similar deposits in South Sweden and Denmark, is to be explained by assuming that, at an early stage in the Late-glacial retreat of the ice, the maximum summer temperature was satisfactorily low, but, with a rapid climatic improvement in the Allerød period (Zone II), the summer temperatures soon became too high, and the plants were eliminated. The modern Scandinavian mountain populations of these plants are presumably therefore derived from per-glacial survivors on coastal nunataks and not from re-immigrants from the south and east. Among the species he discusses are common British mountain plants such as *Dryas* and the least willow (*Salix herbacea*). Dahl suggests that similar patterns may well emerge in the study of the British Late-glacial floras; he points to the abundance of *Salix herbacea* remains in Zones I and II in Ireland (where the maximum summer temperatures were presumably relatively low) in contrast to its apparent absence from the Late-glacial deposits in South-East England. However, we must be very cautious in arguing from negative evidence, particularly when new evidence is accumulating so rapidly; the recent (Godwin, 1954) discovery of the pollen of *Koenigia* in Late-glacial beds in Southern Scotland should offer a healthy corrective to the tendency to argue per-glacial survival from gaps in present distribution, for it seems that even the arctic *Koenigia* may well owe its present reduced area in the British Isles to a process of post-glacial extinction comparable with that which has affected most other members of our mountain flora.

The Irish flora shows examples of two other kinds of distributions, which have been much discussed in this connexion, namely the North American and the so-called 'Lusitanian' types. A discussion of the former is really out of place here, for none of the members of this interesting group has any real claim to be a mountain plant; though the most famous of them, the curious pipe-wort (*Eriocaulon septangulare*), is a feature of the peaty lochs of Skye where many botanists in search of mountain plants can conveniently pay their respects to it. The 'Lusitanian' element in the Irish flora (see p. 59) does, however, present a rather better claim for our consideration. The argument in favour of per-glacial survival for such species does not seem particularly strong, and rests upon the assumption that the climate of Late-glacial and early Boreal times was insufficiently mild along the Atlantic coast of Europe to permit the northward spread of such species to reach their present Irish stations before the Irish Sea was formed. A recent discussion by Professor J. Heslop-Harrison of both these cases will be found in the 1952 Conference Report of the Botanical Society of the British Isles: 'The Changing Flora of Britain'.

The last category of plants worthy of consideration as per-glacial survivors are certain species which appear to be present in more than one form in this country, and for which it is possible to postulate that part of the present-day population is derived from the Continent by re-immigration, whilst part may be derived from isolated western or northern 'nunatak' refuges. Such a situation has been recently investigated by Dr. C. D. Pigott (1954) for the common wild thyme, (*Thymus drucei*). He suggests that the coastal and mountain race of thyme in this country, which is essentially similar to the race in the Faeroes, Iceland, Greenland and W. Norway, survived the last glaciation 'in limited areas around the coasts of the North Atlantic' and spread thence in the Late-glacial flora, whilst the chalk grassland race of southern England, which resembles the common plant in North-West France, has had a separate history of re-immigration from the south, and its main spread has been much more recent, probably in Neolithic times.

An interesting case which appears to be somewhat comparable, though a good deal more complex, is now under investigation by Mr. K. M. Goodway; this concerns the slender bedstraw (*Galium pumilum*), the northern form of which in Britain closely resembles the Icelandic plant and differs markedly from southern forms such as are to be found on the English chalk (Preliminary note, see Goodway, 1955). In both these cases, it is worth noting that Icelandic botanists do not hesitate to ascribe the presence of a similar form of the plant in Iceland to per-glacial survival—indeed they have little choice, for the evidence is that Iceland has not been joined to any mainland since Tertiary times, and the alternative hypothesis of long-distance dispersal across hundreds of miles of sea does not seem likely for many Icelandic species.

To sum up, we could say that a verdict of 'not proven' must at present be recorded for most cases of per-glacial survival claimed for species in the British flora, but that the type of investigation conducted by Pigott certainly suggests that a good prima facie case can be made for some species having had a complex history of survival and re-immigration in the British Isles. A group on which further investigation would be particularly rewarding is the arctic and alpine sandworts already mentioned (p. 64); for the seeds of '*Arenaria ciliata* agg.' occur not infrequently in Late-glacial deposits, and although it is not yet possible to identify such seeds more closely, it is reasonable to hope that with careful study it may be possible to trace the history of each of the three species of this aggregate now growing in such

M. C. F. Proctor

Plate V. Bog sandwort, *Minuartia stricta*. Teesdale

R. Atkinson

Plate VI. Vernal sandwort, *Minuartia verna*. Penyghent, Yorkshire restricted areas in the British Isles. As Nordhagen (1932) has shown in his studies of arctic poppies (*Papaver radicatum* agg.) in Norway, experimental taxonomic study of relict mountain species can give us many clues to the further elucidation of their history; combined with quaternary research it should provide a great deal of information in the near future.

Part Two

REGIONAL

CHAPTER 5

SNOWDONIA AND THE ENGLISH LAKE DISTRICT (J.R.)

It is logical to combine together into one chapter some description of the mountain flora of Snowdonia and of the English Lake District. Not only are they geographically quite close, but geologically also they are rather similar; and this basic similarity in age and origin, combined with rough correspondence in climate and altitude, is probably responsible for their general vegetational resemblances.

In contrast to the main mass of the Scottish Highlands, which is composed of very ancient metamorphic rocks with great areas of granite intrusions, the Lake District and North Wales mountains consist of rocks of Ordovician and Silurian age, together with igneous rocks indicative of old volcanic activity. The impressive, craggy mountains in both areas (e.g. Snowdon, Helvellyn) are very largely composed of these hard volcanic rocks, whilst the more rounded 'moels' in North Wales or, for example, Skiddaw in the Lake District, are composed of somewhat softer shales and slates. On the whole, the bed-rock in most parts is by no means poor in bases, though the volcanic rocks which may be strongly basic usually weather slowly and release these bases rather sparingly; and partly because of this relatively high base-status as compared, for example, with the Cairngorms, grasslands rather than the heather, bilberry or crowberry communities occupy most of the mountain slopes, particularly in the Lake District. These mountain grasslands, dominated by bent-grass (*Agrostis* spp.) and fescues (*Festuca* spp.) or in the badly-drained places by mat-grass (*Nardus*), owe their present extent undoubtedly in part to sheep-grazing, which has depressed or often eliminated the natural forest belt on the lower mountain slopes (see Ch. 2, p. 33).

Even where moorland dominated by heather, bilberry or crowberry is well developed—as, for instance, on parts of the Glyder range in

North Wales—it is not usually possible to find the clear zonation so readily observed on the Cairngorms (see Ch. 6, pp. 113-4). One of the few areas where I personally have seen such an altitudinal zonation in Snowdonia is on the east-facing slopes of Y Garn, above Cwm Idwal. Here dominant heather gives way, on a steep, boulder-strewn slope at about 2,200 feet, to a zone in which bilberry and crowberry are co-dominant, and that in turn, as one ascends further, gives place on the summit ridge (at about 2,800 feet) to a wind-eroded open community containing a great deal of the moss *Rhacomitrium* and the least willow (*Salix herbacea*). It seems possible that in the rather wet climate of Snowdonia and the Lake District such a zonation develops well only where there is hard dry acid rock such as is provided by stabilised scree.

In addition to grassland and moorland types of mountain vegetation, there are in both these districts considerable areas of bog communities, which are usually dominated by bog-moss (*Sphagnum*) or cotton-grass (*Eriophorum*). These bogs may occupy large areas near the heads of mountain valleys or smaller areas on the mountain tops where the drainage is impeded. Such areas have a poor flora; sundews (*Drosera* spp.) and the lovely bog asphodel (*Narthecium ossifragum*) are among the few flowering plants of distinction to relieve their monotony.

From the point of view of the botanist who likes to see rare plants neither Snowdonia nor the Lake District is especially rewarding. If we exclude the two critical genera, the hawkweeds (*Hieracium*) and the eyebrights (*Euphrasia*), there is only one mountain plant to be found in each district which is not also found, usually in much greater profusion, in the Highlands of Scotland—namely *Lloydia serotina* (Pl. 9, p. 116) in Snowdonia and shrubby cinquefoil (*Potentilla fruticosa*) (Pl. 10, p. 117) in the Lake District. A long account of the species which occur in either area would therefore, in the present context at least, be a mere sop to those who are so unlucky as never to visit the Highlands.

Some reference has already been made to Johnson's pioneer exploration of Widhfa, alias Snowdon. Whichever mountain in Snowdonia he in fact climbed, he certainly succeeded in discovering a commendable proportion of the rarer mountain plants of the district. A modern botanist paying his first visit to Snowdonia, if he wished to achieve the same results as Johnson, might well find that he had to make two, if not indeed three, different excursions. For northern rockcress (*Cardaminopsis petraea*) he would go, I expect, to the great cliff, on the way up to Snowdon itself from Llanberis, called Clogwyn Du'r Arddu; and if he were as idle as at least one botanist I know, he would take the mountain railway to the station which is wisely called Clogwyn for short. For alpine saxifrage (*Saxifraga nivalis*) he would climb up, this time necessarily on foot, into the upper level of Cwm Glas. And for any other plants on Johnson's list which he had not succeeded in seeing when looking for these two he would be well advised to try the cliffs around Cwm Idwal. Doubtless certain local botanists, especially Mr. Evan Roberts of Capel Curig, would contrive to show him everything that Johnson saw, and more, in a single day's walk. But to the uninitiated Snowdonia is a formidably large area, and without expert guidance Johnson's achievement would not be very easily repeated today. Owing to its accessibility Snowdon has suffered over the centuries more than most British mountains from the depredations of collectors.

Among the mountain plants of Snowdonia which Johnson did not record, perhaps the most interesting, apart from *Lloydia* itself, are the two chickweeds, *Cerastium alpinum* and *C. edmonstonii,* which are to be sought in the localities already named; mountain avens (*Dryas octopetala*), which inhabits in Wales only two small outcrops of limestone, one on each side of the Nant Ffrancon pass; the hawkweed, *Hieracium holosericeum,* which was apparently familiar as a Snowdonian plant even to John Ray himself, probably from the plateau above Twll Du, but is in these days known in Wales, to me at least, only on the Carnedds; the holly fern (*Polystichum lonchitis*), which in England and Wales alike is now—and this time thanks mainly, I suspect, to gardeners—hovering on the edge of extinction; and finally, and most interesting of all, the two species of *Woodsia.* I deliberately omit both alpine penny-cress (*Thlaspi alpestre*) and the two spleenworts, *Asplenium septentrionale* and *A. x breynii,* because none of the three, in Wales at any rate, is a true mountain plant. The first two grow together, for instance, on the debris of a disused lead-mine not far from Bettws y Coed, at an altitude of only 250 feet, while the only Snowdonian station I know for the third, where it is accompanied of course by its parents *A. septentrionale* and *A. trichomanes,* is also well down below the 1,000 foot contour. I believe that what *Thlaspi alpestre* and *Asplenium septentrionale* demand is not so much a particular altitude, nor even a particular climate, as a particular mineral in the soil. Anyhow it is remarkable how often each of them grows around lead, copper or other mines.

In the case of *Woodsia ilvensis*, on the other hand, which in its best known Snowdonian locality grows almost at the entrance to an old asbestos mine, the proximity is, I imagine, if not a mere coincidence, at least a fact of less significance; for I know of no further evidence, as I do in the case of the other minerals, to suggest that asbestos has any noticeable effect upon the vegetation.

The most celebrated station for *Lloydia serotina* is the Devil's Kitchen itself, where, as is so often the case with rare plants, it is effectively camouflaged by a profusion of a superficially similar flower, the common wood anemone (*Anemone nemorosa*). The Devil's Kitchen is well named, and I wish anybody luck who tries to find *Lloydia* in flower there. I have seen it there myself, and in fine flower too, on one of the dirtiest June days I have ever experienced; not only was it raining in torrents, but there was a full gale blowing as well. I was led to the spot by a very kind guide; otherwise on such a day I should never have got there. Shaking like a leaf from the cold, I stood with my feet in the pool at the top of the great chasm and scanned the cliffs around through binoculars. There the flowers were, sure enough, but since they were shaking in the gale even more violently than I was, I cannot honestly claim to have had a good view; and I admit with shame that I have never had a better, for at the last opportunity I was misguidedly preoccupied with hawkweeds. I do not believe, as a matter of fact, that even the most expert climber could ever reach the particular plants in this colony which flower; otherwise I should hesitate to be so precise about their locality. But there are several other places in Snowdonia where *Lloydia* has been recorded, and patient search may well reveal others yet. It is certainly an unusually romantic plant to seek, both because of the grandeur of its very restricted British territory, and even more because of its world distribution. It occurs not only in the mountains of northern Europe, Asia and America, but also in the highest ranges far to the south, including both the Alps and the Himalayas. Why it should be confined in Britain to Snowdonia, when so many mountains in Scotland would seem at least as suitable, is a mystery; but at least it eliminates any tendency to regard the mountain flora of Snowdon as nothing but a feeble reproduction of that of the Highlands.

What *Lloydia* does for Snowdonia, the shrubby cinquefoil does (if less romantically, because it occurs elsewhere in England and in Ireland) for the Lake District; and if a botanist is in doubt which areas of Lakeland will best repay exploration, he cannot do better than to resort to the recorded stations of the cinquefoil. On and around the Red Screes of Helvellyn it has not, I believe, been seen for a good many years; but there is every reason to suppose that it is somewhere still lingering on. This is an area which very few botanists, and they only the most agile, can claim to have explored at all thoroughly; and at least the search for the cinquefoil there, even if finally unsuccessful, will reveal most of the other mountain species of the district, including certainly mountain avens, the sedge *Carex atrata* and the alpine meadow-grass (*Poa alpina*), and possibly also the rare grass *Phleum commutatum*, and the equally rare mountain bladder-fern (*Cystopteris montana*). On the screes above Wastwater, on the other hand, the cinquefoil has lately been rediscovered, after an interval of many years, by Mr. Derek Ratcliffe, who has very kindly sent us the following notes:—"The habitat is a typical example of the calcicole mountain plant localities on the Scafell Range. The rock here is mostly hard and acidic volcanic material, and calcareous outcrops occur mostly in deep ravines—these apparently lie along fault zones where there has been considerable impregnation with the so-called "iron-ore" and calcite. The resulting extremely rotten and reddish-coloured rock is easily eroded by streams, and so great ravines are developed. There are extensive outcrops of this rotten rock on Wastdale Screes, and especially in the gully where the cinquefoil grows.

"The actual habitat consists of rather wet and often crumbling ledges of this calcareous material, frequently with a good deal of "mull" type humus and with a fairly dense growth of other, smaller plants. In places, small plants of *Potentilla fruticosa* were sprouting from tiny crevices in roseroot fashion. The plant occurred in fair abundance (i.e. scores of scattered plants) from about 1,300—1,500 feet. Some plants grew up to 30 inches, but there were no individual clumps quite as large as the best ones on the Pillar, where its distribution is more restricted than on the Screes.

"Associated species (i.e. growing within a few feet) were *Sedum rosea*, *Oxyria digyna*, *Thalictrum minus* agg., *Saxifraga oppositifolia*, *S. aizoides*, *S. hypnoides*, *Alchemilla alpina*, *Selaginella*, *Festuca vivipara*, *Geum rivale* and *Filipendula ulmaria*. . . . *Asplenium septentrionale* grows in another part of the same ravine. . . . The flora is very interesting, and includes plants such as *Osmunda*, *Aquilegia* and *Teesdalia*, besides the definitely montane species. One plant recorded from the place which I have so far failed to find is *Cardaminopsis petraea*—also recorded from Scafell."

In the locality on the Pillar to which Mr. Ratcliffe refers, the cinquefoil is again accompanied by a number of interesting mountain species, including alpine saw-wort (*Saussurea alpina*). The following full list of the plants closely associated with the cinquefoil on the Pillar was compiled by Max Walters when he and I visited the locality on August 16th, 1953. Incidentally the cinquefoil itself was at that late date in such full flower that it could be seen from at least a hundred yards away; and since the locality is near a well-trodden path, we were astonished that it should be so little known as apparently it is.

Agrostis canina	(bent)
Alchemilla alpina	alpine lady's mantle
A. wichurae	(lady's mantle)
Anemone nemorosa	wood anemone
Angelica sylvestris	angelica
Anthoxanthum odoratum	sweet vernal grass
Campanula rotundifolia	harebell
Carex cf. *lepidocarpa* / *C. pulicaris*	(sedges)
Crepis paludosa	marsh hawk's-beard
Deschampsia caespitosa	tufted hair-grass
Euphrasia rivularis	(eyebright)
Festuca rubra / *F. vivipara*	(fescues)
Galium boreale	northern bedstraw
Geum rivale	water avens
Heracleum sphondylium	hogweed
Hieracium ampliatum / *H. anglicum*	(hawkweeds)
Luzula sylvatica	greater wood-rush
Oxyria digyna	mountain sorrel
Pinguicula vulgaris	butterwort
Saxifraga aizoides	yellow mountain saxifrage
S. hypnoides	mossy saxifrage
S. stellaris	starry saxifrage
Sedum rosea	roseroot
Solidago virgaurea	golden rod

Succisa pratensis	devil's bit scabious
Thalictrum alpinum	alpine meadow-rue
T. minus agg.	lesser meadow-rue
Thymus drucei	wild thyme
Vaccinium myrtillus	bilberry
Viola riviniana	dog violet

Here (as also, evidently, on Wastdale screes) there is so widely different a community from that in which the cinquefoil grows on the limestone of either Upper Teesdale (see Pl. 10, p. 117) or the Burren that the question arises which type of habitat is the more characteristic. From the way the plant behaves on the Continent, it appears that both the Teesdale and the Burren localities represent, geologically if not climatically, much what we might expect, while the stations in the Lake District are abnormal. It seems likely, as was suggested in Chapter 4, that these Lake District colonies represent isolated survivals, due probably to the local quality of the rock, from a time when the plant was very much more widespread in Britain. In that case, however, they are vigorous survivals: not only is ripe seed set, but, as the number of healthy young seedlings shows, it evidently germinates easily enough on the spot.

The only other plant of the Lake District to which I propose to devote much space is red alpine catchfly (*Viscaria alpina*), which has long been known both from Coniston Old Man in North Lancashire and on Hobcarton Crag, above the Whinlatter Pass, in Cumberland. Since its only other British locality is high on the hills at the head of Glen Clova, this is another plant the occurrence of which in the Lake District calls for some explanation. It has not to my knowledge been seen on the Old Man for very many years, though whether that is because it is no longer there or simply because nobody has recently looked for it, I do not know. But on Hobcarton Crag it is still tolerably plentiful, and Max Walters and I went to see it there the day after our expedition to the Pillar. Hobcarton Crag, like so many of the Lake District hills, is for the most part floristically very uninteresting. On my first visit, before I had located the catchfly, I explored the greater part of the mountain before I found what I was looking for, and saw in the process nothing more exciting than alpine lady's mantle (*Alchemilla alpina*), which is of course abundant over much of the Lake District, starry saxifrage (*Saxifraga stellaris*) and the three clubmosses, *Lycopodium clavatum,* which is not really a mountain plant at all, *L. alpinum* and *L. selago*. At the second visit, therefore, we wasted no time, but made straight for the two steep gullies, on the sheer sides of which—or, to be precise, on one sheer side in either case, that which faces eastwards—the catchfly grows.

The alpine catchfly is the plant of which Professor Clapham, in the recent *Flora of the British Isles,* writes enigmatically: "Visited by butterflies and automatically self-pollinated." I am glad to be able to provide at least partial confirmation of the first half of this statement. Though by the date of my second visit the flowering season was long past, there emerged from the middle of the very first plant which we approached, not admittedly a butterfly, but the next best thing, a moth.* Indeed the great majority of the plants in this locality would be more easily visited by butterflies than by botanists; for the rock is not only sheer but also excessively rotten. On my first visit, when I was accompanied by a climber of some skill and experience, it was lucky that the large chunk which detached itself at the crucial moment passed harmlessly between his legs. For my own part, feeling that this was not as irrelevant an event as it may sound, I pocketed two or three samples of the rock in the hope of eventually getting them analysed.

Again, I think, it is worth quoting the list which we made of the associated plants:

Agrostis tenuis	(bent)
Cryptogramma crispa	parsley fern
Deschampsia flexuosa	wavy hair-grass
Festuca vivipara	viviparous fescue
Solidago virgaurea	golden-rod
Vaccinium myrtillus	bilberry

Once again this affords a most striking contrast to the community in the plant's other station. On the hill called in the Floras (but, so far as I know, nowhere else) Little Kilrannoch, the most striking, and indeed almost the only, plants closely associated with the catchfly are Scottish scurvy-grass (*Cochlearia micacea*), mossy cyphel (*Cherleria sedoides*) and thrift (*Armeria maritima*), of which the last alone could

* On a later visit, on July 7th, 1955, the catchfly was in full flower. During about half-an-hour's stay, the flowers (a fair proportion of which in this station are white or very pale pink) were visited by one moth, one bumble-bee and several flies.

conceivably occur on Hobcarton. It is a thousand pities that no data are available concerning the third station on Coniston Old Man. But even without that additional information, there is already in the case of the catchfly a much simpler solution to the problem than in that of *Potentilla fruticosa.*

In Scandinavia, I am told, the alpine catchfly regularly grows, as the spleenwort *Asplenium septentrionale* does in Britain, on and around old copper mines. On Little Kilrannoch, on the other hand, it grows on shingly patches, outcropping through the peat, of decomposing serpentine. That is, I believe, a well-known fact; but in any case the analysis of another of my samples duly confirmed it. There were therefore strong *a priori* grounds for expecting that the rottenness of the rock on Hobcarton is due to an unusually high content of some potent mineral, and that the catchfly can survive in Britain only in conditions that are so repugnant to the majority of plants that any severe competition is eliminated. Unfortunately my own samples from Hobcarton proved insufficient, and others that have since been brought me are still awaiting analysis. But the following notes by Mr. Drummond Matthews, who kindly collected these later samples for me, may throw some light on the problem :—

"Three gullies run from the valley floor up the east face of Hobcarton Pike. . . . The rock is very highly contorted Skiddaw Slate. . . . About 200 feet up in the middle gully a quartz vein outcrops in the right-hand wall. It is associated with Pyrites but not otherwise remarkable. At about 500 feet up in the same gully a more massive quartz vein outcrops, again in the right-hand wall. Once again it is associated with Pyrites. A few feet lower down a very similar vein outcrops in the right-hand gully, where it is better exposed. I think it is a faulted continuation of the vein in the middle gully. The vein rock is grey when unweathered, and in hand specimen I can see only Pyrites and Quartz. On weathering it becomes darker and crumbles. A heavy grey "sludge" of quartz (?) fragments is developed, with conspicuous pale yellow-green staining. The slates have been baked and even perhaps slightly metamorphosed at the contact with the intrusion. . . . The vein is certainly unusually rich in Pyrites, and perhaps in other things too, since it was evidently intruded when quite hot. A competent mineralogist seems to be required! " All that I can add at present to these notes is that the points at which this vein outcrops are precisely the points at which the catchfly grows. Though a competent mineralogist might indeed throw further light on the problem, the general lines of the solution seem already therefore tolerably clear.

There are three last topics on which I must just touch before I bring this chapter to a close. First is the parsley fern (*Cryptogramma crispa*), which is so conspicuous and attractive an adornment of the screes both in North Wales and the Lake District that no account of the mountain flora of these areas, however brief, should altogether omit it. It would be easy to conclude, from some of the current Floras, that this was yet another plant to be seen in greater quantity and finer form to the north of the Scottish border than to the south of it. Such an impression, however, is to my mind quite false. Not, of course, that there is any scarcity of the plant in Scotland: on some of the Scottish hills it may even be as frequent as on the Northern English and Welsh. But if you want to see the parsley fern at its greenest and best, then go to the grey screes of Snowdonia or Lakeland. Incidentally, there is a very interesting account of the vegetation of the screes in the Lake District and North Wales by Leach (1930), in which is emphasised the unique rôle of the parsley fern as the only British vascular plant that is virtually restricted to mountain screes.

Next there are the two endemic eyebrights, *Euphrasia rivularis* and *E. cambrica,* of which the former occurs in both Snowdonia and the Lake District but, so far as is at present known, nowhere else in the world, while the latter is probably, as its name suggests, confined to Wales, if not indeed to Snowdonia, but may possibly also occur in Westmorland. Both are true mountain species, growing chiefly on steep grassy slopes in the hills; in Cwm Idwal, for instance, both are to be found, growing fairly close one to the other, but in conveniently separate colonies. It is a significant point of plant geography, on which some comment has already been made in Chapter 4, that neither of these species should occur in the Scottish hills.

And finally a word about two of the hawkweeds of the Lake District. On August 17th, 1888, the Rev. H. E. Fox, who contributed as much as any botanist to our knowledge of the flora of the Lake District mountains, discovered for the first time in Britain two species of hawkweed which Pugsley in his *Prodromus* has since identified with the two Scandinavian species *Hieracium praetenerum* and *H. sparsifolium.* He found, evidently, a mere two plants of each species, and he labelled the specimens which he collected with the words: "In a ghyll on the back of Kirk Fell, Ennerdale, at 2,000 feet." Since that day neither species, apart from a dubious specimen of the former from Aberfeldy in Perthshire, had ever to my knowledge been seen again in Britain until the summer of 1953, when I determined to try to rediscover them in Fox's original station. Unfortunately his label contains a certain ambiguity: do the words "on the back of Kirk Fell, Ennerdale" mean on the side overlooking Ennerdale, in which case "Ennerdale" is added as an explanatory afterthought, or do they indicate rather that Fox regarded Kirk Fell as being itself in Ennerdale and therefore thought of its back as the side towards Wastdale? The probable answer, of course (which however did not occur to me at the time, because I did not know that there are two Kirk Fells in the district) is that the word "Ennerdale" is added merely to distinguish the right mountain from the other of the same name overlooking the foot of the Whinlatter. But even that gets us little further, because the question remains, which side Fox regarded as the back. Anyhow I guessed wrong, and I spent two whole days (for which, however, I was rewarded by the discovery of the lovely *H. holosericeum* in a new locality) on a false trail. But the third day not only brought me, by elimination, to the right ghyll, it also revealed both the objects of my search; and a very thorough exploration of the whole ghyll showed that, while there are now three plants of *H. praetenerum*, of *H. sparsifolium* there are still only two. In all my days on the hills I do not think that anything has struck me as more remarkable than that these two very rare plants, in the passage of sixty-five years, should neither have disappeared nor appreciably increased, but should rather have remained almost exactly as they were. When, as is probably the case here, a plant is a diminishing species, the actual process of diminution is evidently sometimes very protracted.

CHAPTER 6

THE CAIRNGORMS AND LOCHNAGAR

(*a*) Vegetation (M.W.)

The largest area of continuous high ground over 2,000 feet in the British Isles is the ancient dissected plateau of the Cairngorms in the Eastern Highlands of Scotland, where there are some 150 square miles of uninhabited mountain country traversed only by tracks. Four of the Cairngorm summits—Cairngorm itself, Ben MacDhui, Braeriach and Cairntoul—exceed 4,000 feet, and several other tops fall little short of that height. The individual mountains are not spectacular; indeed Cairngorm well illustrates the fact, familiar to hill-walkers, that height above sea-level is itself of relatively small importance in determining how impressive a mountain may be, for it is from all sides a singularly dull eminence, and its ascent from Glenmore Lodge a simple—and rather boring—gentle uphill walk. Scenically, the Cairngorms are redeemed by the fine corries scooped out of the main mountain mass during the glaciations; the enormous corries to the west of Cairngorm summit provide impressive features of the skyline seen from the north, and carry well into the summer every year snow-fields visible for many miles. This skyline is broken in the middle by the famous steep-sided Lairig Ghru pass, which rises itself to over 2,700 feet, and into which opens Garbh Coire of Braeriach, where a large 'permanent' snow-patch feeds the infant River Dee.

Geologically, the Cairngorms are very uniform, the high plateau being composed of a rather coarsely-weathering granite, which yields a freely-drained acid soil. The poverty of the rock in available bases is no doubt largely responsible for the predominantly moorland type of vegetation over the whole area; whilst the extreme permeability—coupled also with the relatively low rainfall of the Eastern Highlands—explains why large areas of peat bog are rather infrequent. As we shall see in detail later in this chapter, the acidic rock has a marked

Photographs by M. C. F. Proctor

Plate VIIa. (*left*). A mountain grass, *Phleum commutatum.* Clova Mountains
b. (*right*). A mountain sedge, *Carex capillaris.* Clova Mountains

M. C. F. Proctor

Plate VIII. A mountain chickweed, *Cerastium cerastoides*. Ben Nevis.

effect on the flora—many mountain plants common on some other Scottish hills occur in the Cairngorms only in very restricted areas where locally limestones or other basic rocks are found.

Scenically and floristically, therefore, this great mountain mass is apt to be somewhat disappointing, at least on a first, unguided visit. The vegetation as a whole, however, presents features of very considerable interest to any serious student of British mountain botany, as has been revealed by the studies, based upon expeditions in 1938 and 1939 from the University Botany School, Cambridge, published in the Journal of Ecology (Watt and Jones, 1948; Metcalfe, 1950, Burges, 1951). Moreover, the recent acquisition by the Nature Conservancy of a large area of the Cairngorms as a National Nature Reserve should mean that in the near future even more information will be forthcoming on the many problems of mountain vegetation which can be studied on this great mountain range.

It seems very probable that the Cairngorms present a mountain vegetation essentially unaffected directly or even indirectly by human activities—and, as we have already seen in Chapter 2, it is very doubtful whether this claim could be made for many other familiar mountain masses in the British Isles. At the present time, deer-grazing is the only interference factor of any importance on the Cairngorm mountains, and though sheep and cattle were grazed locally in the area in past times, it seems most unlikely that either had much effect at high levels. The tree-limit (native pine and birch) in the area is, of course, almost certainly artificially low through human interference, and the rather abrupt transition from pine-birch woodland to the moorland communities is hardly a natural feature; but above the tree-limit, the vegetation is natural and relatively undisturbed.

The comparative freedom from disturbance of the Cairngorms mountain vegetation makes it of particular interest that we should find it possible to recognise here a rough zonation of communities, the upper zones of which at any rate correspond remarkably well with those found on many Scandinavian mountains (fig. 2, p. 34). Thus, above the forest there is a belt in which heather (*Calluna*) is typically dominant, but this is replaced at varying altitudes around 3,000 feet by a belt in which the two common plants are the crowberry *Empetrum hermaphroditum,* and the bilberry (*Vaccinium myrtillus*); and this in turn gives way on the higher exposed ridges and the summit plateaux to open communities in which the moss *Rhacomitrium lanuginosum,* the sedge

Carex bigelowii or the rush *Juncus trifidus* are usually the common plants. This broad altitudinal zonation—much affected, of course, by local variation in topography and therefore in exposure factors—bears a striking resemblance to that on many comparable Norwegian mountains, even to the almost exact correspondence in the species of plants concerned. No higher plants occur commonly on the Cairngorms which are not to be found in remarkably similar situations in Scandinavia—though of course the floristic poverty of the granite is such that the reverse is obviously untrue.

A comparison of the Cairngorms zonation with that developed on acidic mountains further west in the Highlands suggests that such a zonation is perhaps at its clearest in Britain in the rather extreme climate and soil conditions of the Eastern Highlands. Not only is there a general tendency for blanket bog peat formation in the rainier west—which would obviously obscure a moorland zonation—but also there seems to be a telescoping of altitudinal zonation, by which mountain plants in general in the Western Highlands occur at lower altitudes. For both these reasons, perhaps, the crowberry-bilberry zone has not been clearly recognised elsewhere on British mountains, though it must occur widely in a less clearly-demarcated way, as we saw in Chapter 5.

The lowest of the mountain communities—the heather-dominated belt—shows locally an abundance of the bearberry, *Arctostaphylos uva-ursi,* a pretty trailing dwarf shrub with neat evergreen leaves and red berries, which thrives particularly on the loose stony edges of the well-developed 'terraces' (see Ch. 2, p. 39). Another plant common in the heather zone on Cairngorm, particularly where there is half-buried scree or rock debris, is the clubmoss *Lycopodium annotinum* (Pl. 11, p. 124). These two plants seem to achieve a greater importance in the vegetation on the Cairngorms than in many other mountain areas, and they are plants with a marked preference for well-drained stony ground, though the bearberry, unlike the clubmoss, seems to be rather indifferent to the base status of the soil.

As in the heather zone, terraces influence the patterns of vegetation in the crowberry-bilberry zone, mat-grass (*Nardus*) being a common dominant of the damper ground just below the terrace step. Late snow-lie is also of more importance and may locally produce a snow-patch vegetation from which crowberry and bilberry are absent (see below). At the other extreme, too much exposure favours the dominance of *Rhacomitrium*; whilst over most of the summit plateau, where the wind exposure is such that no snow can lie as a continuous cover, and the smaller rock and soil particles are prevented from accumulating, there is a mere sprinkling of tufts of a few hardy plants, chief among which is *Juncus trifidus*. As we shall see later, one or two of the most interesting Cairngorms plants manage to grow on the summit plateau and one, *Luzula arcuata,* positively prefers it to any more sheltered habitat.

The effects of late snow-lie become very marked on the Cairngorms above 3,000 feet. A visitor to these mountains in June will usually find quite considerable patches of snow in certain areas, chiefly in hollows facing north or north-east; and, although the rate of snow melt during the summer is very variable from year to year, a few snow-patches achieve an individuality—and even a local name—and are to be found in the same position, though varying naturally in size, every summer. Such is the conspicuous 'snow wreath' which lies like a garland around the head of the shallow Coire Cas, west of the summit of Cairngorm; but this and several others on Cairngorm normally melt relatively early, and are usually entirely gone by the end of July. In two places in the Cairngorms, however, there are virtually permanent snow patches; one, which has already been mentioned, is in the great corrie of Braeriach, facing north-east over the Lairig Ghru pass, and the other is the picturesquely named Ciste Mhaireraid (Margaret's Coffin), in a small high corrie about a mile north of Cairngorm summit and facing roughly east.

A long series of botanical observations is not yet available for any British snow patch vegetation, but some study of the Cairngorm snow patches does at least reveal that there is a general pattern, or zonation, to be recognised. A snow cover lasting into June or July would seem to prevent a closed community of higher plants from developing at all, and in its place is a more or less continuous carpet of mosses, with occasional isolated tufts of upright fir clubmoss (*Lycopodium selago*), starry saxifrage (*Saxifraga stellaris*), wavy hair grass (*Deschampsia flexuosa*) and the sedge *Carex bigelowii*. Inside the moss zone, it may be possible to recognise an even more lowly liverwort zone, which must normally survive nine or ten months of continuous snow cover and often fail to emerge snow-free at all for a whole season. The whole question of the ability to survive under snow, apparently possessed to a remarkable extent by the snow-patch bryophytes, is one which would be well worth more investigation. During observations of the Braeriach snow-patch over 40 years, Seton Gordon has recorded (Gordon, 1950) only three occasions on which he has seen really *bare* ground at the edge of the melting snow; and Dr. A. S. Watt found, on visiting Margaret's Coffin on the thirteenth of September, 1948, when the snow was reduced to two small patches, neither more than 18 inches thick, that mosses were carpeting the ground at the edge of the remaining snow.

It is not easy to decide, on the basis of the lists from the Cairngorms snow patches, which flowering plant holds the record for greatest tolerance of snow cover. Seton Gordon (loc. cit.) mentions non-flowering rosettes of the starry saxifrage as being nearest the snow in his observations; but Watt's notes for Margaret's Coffin show that the recently uncovered moss zone in September, 1949, had isolated tufts of *Carex bigelowii* and three grasses—wavy hair-grass, mat-grass (*Nardus stricta*) and alpine meadow-grass (*Poa alpina*).

It is interesting, and perhaps rather surprising, that most of the flowering plants of the Cairngorm summits *can* occur at least as individuals under conditions of late snow-lie. Thus at Margaret's Coffin on the 30th June, 1939, I recorded that *Juncus trifidus* tussocks occurred in an open community with much bare gravel right up to the edge of the snow—though there was a gradual, regular decrease in size of the scattered tussocks as one approached the snow. (At that time, incidentally, the snow patch was some 150 yards long and must have been several yards thick at its centre). There are, however, two species which on Cairngorm at any rate are especially closely associated with snow patch conditions—namely least cudweed (*Gnaphalium supinum*) and alpine lady's mantle (*Alchemilla alpina*). The former seems to be rather strictly confined to snow patch habitats, occurring outside them only in small patches or even as isolated individuals on the gravelly sides of mountain streams; it seems to be particularly abundant where the melt-water from the snow keeps the ground wet, and in such situations it is admirably adapted to resist periodic flooding with its silvery 'unwettable' hairiness. The lady's mantle shows a similar adaptation—the young, unfolding leaves have a silvery sheen of hairs on the exposed under surfaces—but its place in the snow-patch vegetation would seem to be characteristically the sloping sides of the hollow in which the patch is situated rather than the very wet seepage areas, although it does also occur there mixed with the cudweed.

M. C. F. Proctor

Plate 9. Lloydia serotina. Snowdonia

R. Atkinson

Plate 10. Shrubby cinquefoil, *Potentilla fruticosa*, on shingle by River Tees

Neither of these plants, it should be noted, shows the same restriction at high levels to snow patches in many other parts of the Highlands; certainly not the lady's mantle, which on the base-rich mountains may be an abundant plant of the widespread grassland. In this as in other respects it seems that the granite Cairngorms exhibit an extreme situation, in which mountain species common in other regions are remarkably local and specialised in their requirements. One of the requirements which may explain the occurrence of certain species on the Cairngorms only in snow-patch vegetation is a suitable rooting medium; for, as we have seen, the granite weathers into a coarse-particled gravel, and it is only under rather specialised conditions such as the snow-patch that finer soil particles can accumulate to any extent.

(*b*) Flora (J.R.)

Although the Cairngorms as a whole are floristically far inferior to many other areas of the Scottish Highlands, yet a complete list of the species found on the hills bounded on the south by the Dee and on the west by the Feshie contains a surprisingly large total of mountain plants. The explanation is simple enough: that this great area of upland, in spite of the floristic monotony of the greater part of it, is none the less sufficiently varied to provide a suitable habitat for mountain plants of three different types. There are in the first place several species, such as rock whitlow grass (*Draba rupestris*), alpine milk-vetch (*Astragalus alpinus*), mountain avens (*Dryas octopetala*) or reticulate willow (*Salix reticulata*), which are characteristic, in Britain at least, of the richer basic rocks. Then there are numerous relatively wide-spread plants, such as alpine meadow-rue (*Thalictrum alpinum*), mountain scurvy-grass (*Cochlearia alpina*), moss campion (*Silene acaulis*), starry saxifrage (*Saxifraga stellaris*), alpine willow-herb (*Epilobium anagallidifolium*), dwarf cudweed (*Gnaphalium supinum*), mountain sorrel (*Oxyria digyna*), least willow (*Salix herbacea*), three-leaved rush (*Juncus trifidus*), spiked woodrush (*Luzula spicata*) or the sedge *Carex bigelowii*, which seem, so long as climatic conditions are suitable, to be more or less indifferent to the underlying rock. And finally there is a comparatively small number of species which are characteristic of acid rocks at high altitudes—the chickweed *Cerastium cerastoides*, for instance, brook saxifrage (*S. rivularis*), curved woodrush (*Luzula arcuata*), the sedge *Carex lachenalii* and the grass *Poa flexuosa*. It is hardly surprising that, with these three separable ingredients, the list of the mountain plants of the Cairngorms reaches a respectable length.

In this instance, however, a bare list is more than usually misleading. The first group of plants in particular, those which are characteristic of basic rocks, are for that very reason not at all characteristic of the central Cairngorm massif. It happens that in two particular areas near the edges of the district, one an isolated hill to the south-east of the main mass and the other a corrie or two off Glen Feshie to the west, there are small but obvious outcrops of limestone or mica-schist; and these limited localities carry a flora representative rather of the Clova hills than of the Cairngorms proper. On the hills around the Lairig Ghru the flora is everywhere (or very nearly everywhere) that of an acid mountain. I have never had any great affection for the broad expanses of the Cairngorms, which seem to me, compared with their sisters in the west, shapeless and dreary lumps; and in consequence I have never been nearer the Glen Feshie corries than the cliffs around Loch Einich, where the most interesting plants are alpine speedwell (*Veronica alpina*), a great variety of beautiful "alpine" hawkweeds (see Note at end of chapter), the willow *Salix myrsinites*, and the grass *Phleum commutatum*. But I have visited, and that more than once, the relatively low hill to the south-east, and it is a place that I would gladly visit yet again. Apart from mountain avens and alpine milk-vetch, each locally abundant, and the latter, on sunny days at least, blatantly flaunting its colony of the very rare mountain burnet moth, it is the best locality I know both for arctic mouse-ear chickweed (*Cerastium edmonstonii*), surely the loveliest British representative of its genus, and, in a slightly different spot from the rest, for the winter-green *Ramischia secunda*.

As for the second element, consisting of plants which seem to be equally at home on basic or on acid soils, this is in other respects by no means a homogeneous group. Whereas, to begin with, the majority are to be found on the hills not only of Scotland but also of England or Wales or both, there are two at least, the least cudweed and the three-leaved rush, which seem to be confined in Britain to the Scottish mountains. Again, while some of those I have mentioned, notably the least willow, the three-leaved rush and the sedge *Carex bigelowii*, are primarily plants of the summits, others, such as the alpine meadow-rue, alpine scurvy-grass and alpine willow-herb show a special predilection for flushes and rills below—sometimes indeed, as in the Isle of Harris, very far below—the tops of the hills. But in spite of these and other such differences, all these species are alike in this one respect, that nobody who has even a slight acquaintance with the mountain flora of Britain as a whole would dream of regarding any of them as exclusively, or even especially, characteristic of the acid mountains of the north-east.

It is therefore with the third group that this section is chiefly concerned, those plants which are most at home, in Britain at least, on acid rocks at high altitudes; and since from this particular standpoint Lochnagar is to be regarded as belonging rather to the granite hills to the north of the Dee than to the mica-schist hills to the south, the list of such plants which are to be found on the Cairngorms must be lengthened by certain others which occur on the cliffs or summit plateau of Lochnagar. Of all the floristically dull mountains in Britain —and to my mind every acid mountain is floristically dull—Lochnagar is surely the most rewarding. True, the approach to the mountain, whether from Deeside, from Loch Muick or from Glen Callater, presents a botanical monotony that would be hard to beat. But the screes, the gullies and the ledges of the two great north-facing corries can boast not only brook saxifrage, alpine speedwell and the grass *Poa flexuosa,* but also blue sow-thistle (*Cicerbita alpina*), highland cudweed (*Gnaphalium norvegicum*) and the hybrid grass *Poa x jemtlandica*; while, on the boggy summit plateau, besides the parent *Carex lachenalii,* there grows also its hybrid *Carex x helvola,* which is accompanied in this, its only known British station, by another rare sedge, *C. rariflora.* For a dull mountain that is an undeniably fine assortment of rarities; nor is it intended as an exhaustive list. I mention these plants in particular because they are as characteristic of the rock upon which they grow as are their counterparts on the other side of the Dee.

The chickweed *Cerastium cerastoides* (Pl. VIII, p. 113), the first of these acid-loving mountain species, is a plant over which it would be hard to wax enthusiastic at first sight: it is too obviously related to certain vulgar weeds. But for all that, its trailing habit and the slender pedicels on which it carries its usually twin flowers give it a grace which is lacking to its commoner relatives. It is to be sought, in my limited experience, either on fine screes or on the grassy slopes above them. In such habitats I have seen it, apart from the Cairngorms themselves, both on Ben Nevis and its neighbours and on the hills around the head of Glen Affric. In each of these localities, moreover, it grows no great distance away from other plants of the same class, on the Nevis range the saxifrage *S. rivularis* (and also its relative, *S. caespitosa*), above Glen Affric the highland cudweed and the wood-rush *Luzula arcuata*.

The two saxifrages, though in the one locality mentioned they do grow within a few feet of one another, seem yet to have no very obvious taste in common. Tufted saxifrage (*S. caespitosa*) is about the rarest of all our mountain plants, several of the older records for it being almost certainly due to confusion with other "mossies". I myself have seen it (and very little of it too) only in the one spot, where it grows, somewhat precariously, in the very narrow chinks of an exposed north-facing rock; and in the other locality where I know it is still to be seen, high up on the Cairngorms, its habitat is, I believe, equally exposed. Incidentally, in the one locality where I have seen it, it does not seem fully to develop its petals, so that it sometimes appears at first sight, even when fully in flower, to be already well past flowering. The brook saxifrage, on the other hand, is a plant of the dampest and shadiest places, often in the close neighbourhood of long-lying snow-patches. So, where tufted saxifrage clings to the exposed rock, brook saxifrage grows instead, in a narrow line, at the very foot of the cliff, where there is a shallow but moist and sheltered overhang. Likewise on Lochnagar it chooses damp mossy rocks immediately beside the little rills which trickle down the gullies from the patches of snow at the top. It is not, as a matter of fact, strictly accurate to describe brook saxifrage as exclusively a plant of acid mountains: it has been found on Ben Lawers itself, about the least acid of all British mountains. But I believe all the same that it has a strong preference for a habitat such as those I have described; and, never having succeeded in finding it on Ben Lawers, I prefer to think that it there inhabits a pocket, if not of positively acid, at least of rather peculiar soil.

Neither the highland cudweed nor the blue sow-thistle has ever, to my knowledge, been found on the Cairngorms proper, though I suspect that the former at least may yet turn up there. But on certain ledges on Lochnagar, vertiginous ledges none too easy of access, they can be seen growing side by side; and sometimes there is a ptarmigan's nest there as well, protected by the large base-leaves of the sow-thistle. The flora of these particular ledges is exceptionally luxuriant and contains several such woodland species as wood-sorrel (*Oxalis acetosella*) and greater woodrush (*Luzula sylvatica*). Two of its chief constituents, the red campion (*Melandrium rubrum*) and the rosebay willow-herb (*Chamaenerion angustifolium*), splash the black rock with a colour that can be seen from the other side of the lochan in the corrie. The rosebay, incidentally, is here beyond any question native, and it is probably significant that it is in a neater form than the usual south-country plant and has thicker leaves which are much more obviously crisped or crinkled. The cudweed and the sow-thistle have interesting and somewhat analogous distributions in this country, the latter in particular being the outstanding example of an arctic-alpine species confined in Britain to the Eastern Highlands of Scotland. Besides Lochnagar it inhabits only, and that very locally indeed, certain cliffs or ravines in several of the grand glens to the south—Caenlochan and Canness, Corrie Kander and (in former days at least, though I fear it has now disappeared from its best-known station in a flood) Glen Clova itself. Like the brook saxifrage, it does not seem to be strictly confined to the granite, and it is perhaps misleading to include it in the group of plants under discussion. But again I am fairly confident, from my own knowledge of it, that it is most at home on acid rocks; for in every locality in which I have seen it, and notably on one small ledge in Caenlochan, it rejects the great expanse of rich mica-schist in the neighbourhood in favour of the relatively poor and slabby granite. Since it is evidently also a favourite food of the red deer, and tends therefore to grow only where they cannot reach it, it is, in most of its stations, one of the few British mountain plants which demand both a certain agility and a fair head for heights.

I cannot resist, in this connection, recounting the Strange Episode of the Major's Spectacles. One hot day in the middle of July I was descending alone one of the long scree gullies that drop into Caenlochan Glen from the west. When I was almost immediately below the ledge on which the sow-thistle grows, I noticed, lying on the scree, and somewhat mud-bespattered but otherwise intact, a pair of horn-rimmed glasses. Thinking abstractedly that their owner would never find them there, I picked them up, put them in my pocket and promptly forgot them. When eventually I emptied my pockets, I was at a loss. To take them to the Braemar police seemed too forlorn a hope, and no alternative occurred to me. In the end, cursing myself for not having left them where they lay, I threw them in disgust, along with the refuse from my vasculum, into the waste-paper-basket in my hotel bedroom, and left for Aviemore.

Next day, accompanied by three fellow-botanists (only one of whom, incidentally, and that not the Major, was previously known to me) I paid my first visit to the hawkweeds of Glen Einich. The cliffs around the head of the loch are formidable and treacherous, and the hawkweeds have a bad habit of growing just out of reach. Suddenly, when we were each busy with our own ploys, a gentle voice reached me from along the cliff: "Could somebody come and give me a hand?" Two of us arrived in time; the Major was still hanging on, and we grounded him safely. "What a good thing I wasn't alone," he said calmly. "Last year I got stuck like that on the ledge in Caenlochan where the blue sow-thistle grows. I was very keen to photograph it and scrambled up to it without bothering to think how I should get down again. That time there was nobody to help, and in the end I pitched on to my head on the scree and was hard put to it to get back over Glas Maol to the Devil's Elbow." "You didn't by any chance lose a pair of horn-rimmed spectacles, did you?" I asked, and went on to tell him, somewhat disingenuously, that I had left them in the Invercauld Arms Hotel at Braemar. He wrote to the manager, of course, and was disgusted with the reply that nothing had been seen of his glasses. I never dared to tell him the whole shameful truth.

The blue sow-thistle must unquestionably rank, if not as the most handsome of British mountain plants, at least as the most conspicuous. The remainder of the present list—that is, curved wood-rush (*Luzula arcuata*), the two sedges *Carex lachenalii* and *C. x helvola*, and the two grasses *Poa flexuosa* and *P. x jemtlandica*—must afford, to all but a few enthusiasts, a depressing contrast. *Luzula arcuata* is the plant *par excellence* of the stony summit plateau, only occasionally, as in the corries of Cairngorm itself and on Carn Eige between Glens Affric and Cannich, spilling a little way down the steep and rocky mountain-sides. It can be distinguished at once from its much commoner mountain relative, *L. spicata*, not only by its more compact habit and its paniculate inflorescence with curving peduncles, but also by the lighter and redder brown of its perianth. I am not sure what is its lowest recorded station in Britain: I still have a sneaking feeling (and also a painting of my father's to confirm it) that long ago, before I knew anything whatever of the plants of the Scottish Highlands, I found it on the summit plateau of the Merrick, in Galloway, which only attains a modest 2,764 feet. But since then I have learnt that in most of its localities it is useless to begin looking for it until at least the 3,000 foot contour has been passed, and in consequence my faith in my earlier find has steadily diminished.

Carex lachenalii seems to be essentially a plant of damp north-facing rocky slopes. Such is its habitat on Lochnagar, where it grows in more than one spot not very far below the top of the ridge running from the Stuic to the summit proper; and in its other localities too, Cairngorm or Cairntoul for instance, it frequents very much the same sort of ground. It too, like the sow-thistle, is pre-eminently a plant of the Eastern Highlands; but the fact that it occurs also in at least one of the big north-facing corries along the Ben Nevis range means that it is well worth while looking for it considerably to the west of the Cairngorms and Lochnagar. As for the hybrid between it and *C. curta*—or perhaps I should say, the supposed hybrid between these two, for I do not know that the parentage of *C. x helvola* has ever been experimentally confirmed—it has been found in Britain, so far as I know, only in one or two small boggy hollows on the same ridge of Lochnagar. But since it is by no means a conspicuous or distinctive plant, it seems quite likely that it has so far eluded detection in others of the localities of its rarer parent.

The last two plants on the list, *Poa flexuosa* and *P. x jemtlandica,* are also the two most difficult. I was under the impression until recently that I had seen the former in what is probably its best-known station in Britain, the long scree shoots under the summit of Ben Nevis; and I had certainly seen, in fair quantity, the very undistinguished grass that British botanists have hitherto called by its name. But during the summer of 1953 an excursion of the Botanical Society of the British Isles found, on a ledge in one of the Cairngorm corries, a grass which Dr. Melderis unhesitatingly pronounced to be conspecific with the Scandinavian *P. flexuosa,* but which was apparently not identical with the Ben Nevis plant. There seems to be a problem here which, at the moment of writing, is still awaiting a solution. As for *P. x jemtlandica,* it is supposed, and surely rightly, to be the hybrid of *P. flexuosa* and *P. alpina.* At any rate in its only known British station, the cliffs and gullies of Lochnagar, it is accompanied by both its putative parents. It is also morphologically intermediate between the two, its leaves being shorter, broader and stiffer than those of *P. flexuosa,* and its inflorescence, unlike that of *P. flexuosa,* being viviparous. Finally, it may be significant that, in my experience at least, *P. flexuosa* chooses the open scree, *P. alpina* usually cliff-ledges, and

P. x jemtlandica the boulders or outcrops of rock on the edges, or even in the middle, of scree gullies.

I cannot leave the Cairngorms without at least a mention of a number of other plants which may well be thought, by those who know the district, to be no less characteristic of it than those I have already discussed. First is northern rock-cress (*Cardaminopsis petraea*), which, being frequent in a number of spots high up in the Cairngorms, including the neighbourhood of the Wells of Dee, has been carried far down the river to colonise, and very effectively too, the wide expanses of riverside shingle on either side of Braemar. Of all British mountain plants the northern rock-cress has probably the most puzzling distribution (see Appendix III, Map 3a). Personally I regard it as primarily a plant of volcanic rocks in the west: it is at any rate abundant in such localities as Ben Hiant in Ardnamurchan, Ruinsival in Rhum and both the Cuillins and the Storr in Skye. But since it is also found on the cliffs of Snowdon, the screes of Wastwater, one acid summit in the Isle of Harris, and the Cairngorms, there is no denying that its tastes are unusually catholic. At all events, it is emphatically not one of those plants which are especially characteristic of the North-Eastern Highlands; and with that observation I must leave its distribution map to speak (if it will) for itself.

The next four species are a heterogeneous collection—mountain azalea (*Loiseleuria procumbens*), alpine speedwell, the grass *Deschampsia alpina* and alpine lady-fern (*Athyrium alpestre*). I tend to think of them myself as being more abundant on Lochnagar than anywhere else in Britain. But since all four of them, though all but the grass confined to Scotland, occur there not only on the acid hills but also, if less freely, on the basic—on Ben Lawers itself, for instance, and in Glen Clova—my reasons for regarding them as centred on Lochnagar cannot be altogether valid. They probably are, I think, genuinely calcifuge; but since a calcifuge plant can often, thanks to leaching, find a spot to suit it even on the richest mountains, it is naturally better equipped to spread from the acid hills where it is most at home than is a calcicole plant to spread from the basic on to the acid rock. That is of course one reason why the basic hills are floristically so much richer than the acid.

Last of the plants of the Cairngorms proper comes Boyd's pearlwort (*Sagina boydii*), which constitutes, probably, the greatest mystery in the annals of British botany. Though the story is well known, it may still

J. Markham

Plate 11. Interrupted club-moss, *Lycopodium annotinum.* Cairngorms

J. G. Roger

Plate 12. Diapensia lapponica. Scotland

be worth recounting yet again. I shall tell it mainly in the words of G. C. Druce, taken from his article on "Dubious Plants of Britain" in the *Report of the Botanical Exchange Club* of 1920. *Sagina boydii* was supposed to have been found in 1878 by Mr. W. B. Boyd somewhere in the neighbourhood of Braemar. Most unfortunately, "he does not remember gathering it, but he found it in his potting shed among other specimens he brought back from Braemar, and his impression is that he had gathered it on Ben A'an. But he had received about that time living plants from Switzerland for his garden, and there is the chance that it may have a foreign origin. In later years his memory rather crystallised upon Ben A'an; directly after his reporting it he was less certain. The plant is known from nowhere else. . . . In order to check my remembrance of the details of its occurrence as given me by Mr. Boyd (I had many talks with him about it as I was anxious to refind it), my remarks were submitted to Professor I. B. Balfour, who kindly tells me that Mr. Boyd in the early nineties, on seeing his plant of *S. boydii* at Edinburgh, said: 'I have no recollection of getting that plant. I believe it to be a Braemar plant, but when I was arranging for planting my Braemar collections I had also in hand some Swiss plants, and it may have been one of them.' As Professor Balfour says: 'There is nothing in it—whether it be species or otherwise—to forbid its being a British plant; nothing in it to say nay to a continental origin. Re-finding is what is to be aimed at.'" Unfortunately re-finding is still what is to be aimed at. If Mr. Boyd himself did indeed see it in the field, he is the only botanist who ever has. And if Ben A'an is probably the area to be explored first, an already difficult quest is made even more difficult still by the existence of another tradition to the effect that the plant was found "in a dell in the deer forest of Invercauld". But there is one redeeming feature in the story, that the plant still exists in cultivation. Even if it is never refound, there is at least incontrovertible evidence that somebody, somewhere, did once find it.

There is still one final postscript to be added to this chapter. The blue heath, *Phyllodoce caerulea,* even if long ago it was recorded from Aviemore, is most probably not, strictly speaking, a plant of the Cairngorm massif at all; it belongs rather to that central group of the Grampians of which Ben Alder is the undisputed monarch. None the less the Sow of Atholl (which, as every botanist knows from the Floras, is the only British locality now known for the plant in question) is so conspicuously acid a mountain that, if a place is to be found anywhere in this book for *Phyllodoce caerulea,* it can only be here. Fortunately there is little to say on the subject. Being apparently a rather shy flowerer, *Phyllodoce* is not by any means an easy plant to find. Its locality, a steep and stony slope thickly overgrown with heather, is dismally lacking in any notable or memorable features, and the plant itself, when not in flower, is much too like crowberry (*Empetrum*) to be easily spotted. I myself had to pay three visits to the Sow of Atholl before I achieved my objective, and the only recompense for my first two unsuccessful visits, apart from few-flowered sedge (*Carex pauciflora*) in the bog at the mountain's foot, was mountain azalea and interrupted clubmoss (*Lycopodium annotinum*). But for all that, the plant is there in fair quantity, centred admittedly on one particular spring, but extending, in a narrow belt at the same altitude, for some considerable distance across the face of the hill. As I have already said, the early record from Aviemore seems likely to be either a mistake or else an exceptionally early instance of botanical dissimulation. But to judge from its only known British station on the Sow of Atholl, there is not the faintest reason why *Phyllodoce caerulea* should not even yet win a rightful place among the plants especially characteristic of the Cairngorms proper. It is true that both in the Pyrenees and in the Alps it seems also to have only one known station; it is evidently a plant with a liking for small and isolated colonies. But since that peculiarity is not nearly so noticeable in the north of Europe as in the south, I shall continue to hope that its station on the Sow of Atholl will not for ever remain its only known British locality.

Note on the Hawkweeds found on British Mountains (J.R.)

As I suggested earlier in this chapter, the most spectacular element in the flora of the Cairngorms area as a whole is the profusion and variety of its hawkweeds. Since there is almost certainly no mountain in Britain without its population of hawkweeds, some space must be found for them somewhere in this book; and since, for the truly montane species, the Cairngorms are without question the best of all British hunting-grounds, this is probably the most suitable place to insert such observations as I have to make about them. I must start, however, by narrowing the field. The number of described species of the genus that can be found in Britain at an altitude of 2,000 feet or more amounts, at a very rough calculation, to something between 75 and 100, of which an appreciable proportion—say one third—belong to the various subsections of the one very large and very difficult section *Vulgata*, while another dozen or so fall into the sections *Oreadea* and *Sub-Oreadea.** These whole sections, and likewise the relatively few species of the *Transitoria* and *Aphyllopoda* which are sometimes found at high altitudes, I must in the present context ignore: for it is the species of three sections only, namely *Alpina, Subalpina* and *Cerinthoidea,* that constitute in Britain both the spectacular and the essentially montane element in the genus. There are indeed few British plants more strikingly handsome than these mountain hawkweeds, with their neat basal rosettes, their velvety-black involucres, often with a halo of long white-tipped silken hairs, and their broad golden flower-heads (Pl. X, p. 129). And the way in which each mountain-range in Scotland has its characteristic assortment of species makes some familiarity with the genus, if originally hard to acquire, ultimately a very rewarding possession.

It will be as well to eliminate at the outset one very common cause of confusion and disappointment. Of all the showy large-flowered species of hawkweed in Britain, by far the most widespread and abundant—in Northern England, Scotland and Ireland at least, but mercifully not in Wales—is *H. anglicum*: it is probably no exaggeration to say that at least half the hawkweeds noticed by the average botanist on the Scottish hills would belong to this species. Unfortunately it is no easy matter so to describe *H. anglicum* that it cannot be confused with other and rarer species; indeed it is too variable to be accurately described at all. But in its most usual form it has, all the same, two characteristics which neither the *Alpina* nor the *Subalpina* ever share with it, namely a broad and *semi-amplexicaul* stem-leaf and *pale* yellow (as opposed to golden) florets. It is a useful general rule that no hawkweed of which the lower cauline leaf shows any sign of clasping the stem can belong to either of the sections *Alpina* or *Subalpina.*

Of the section *Alpina* there are no fewer than four species which are at present known nowhere in the world outside the Cairngorms; *H. grovesii, H. macrocarpum, H. graniticolum* and *H. pseudo-petiolatum.* That fact alone at once puts the Cairngorms in a class by themselves among British mountains for the hieraciologist. And apart from these

* Throughout this note I adopt the classification and nomenclature of H. W. Pugsley, *A Prodromus of the British Hieracia* in *Journal of the Linnean Society of London* (Botany) Vol. 54 (1948).

four, each of which involves a long walk with a patient search at the end of it, there are several other species of the group which can be relatively easily found in the district. In Glen Einich, for instance, I have seen in fair quantity seven other species of this one section, in some places growing together but in others in separate colonies. The most frequent, I think, were *H. alpinum* itself (in an exceptionally handsome form), *H. eximium,* including the variety *tenellum,* and *H. backhousei*; but *H. calenduliflorum, H. globosiflorum, H. pseudo-curvatum* and *H. hanburyi* are all there too. The long stretch of cliff around Loch Einich is a particularly savage place, and even the most avid collector could make little impression on the hawkweed population. And so I have no hesitation in saying that for abundance and variety of hawkweeds of the *Alpina* section Glen Einich is by far the richest place I have yet struck.

Other areas of Scotland where the *Alpina* are especially well represented are Lochnagar, where *H. eximium, H. calenduliflorum, H. gracilentum* and *H. hanburyi* are all tolerably frequent; Glen Fee, in Angus, where *H. hanburyi* is much the most abundant, but where *H. eximium* and *H. gracilentum* are again not uncommon; the glens around Clach Leathad, at the head of Glencoe, where there is exactly the same assortment; and, best of all after the Cairngorms, certain spurs to the North of the Nevis range, where the variety is increased by the addition of *H. alpinum* itself (elsewhere in Britain known only in the immediate neighbourhood of the Cairngorms), *H. backhousei* and apparently also *H. subgracilentipes.* It is noteworthy that in most of these areas the rock is markedly acid. My own experience leads me to think that whereas the true *Alpina* show a preference, in Britain at least, for the acid rocks, the majority of the *Subalpina* show an even stronger taste for the mica-schist (see Appendix III, map 4a). At all events there are many Scottish mountains not otherwise notable for the richness of their flora, hills such as Sgurr na Ciche at the head of Loch Nevis or Ben Klibreck in Sutherland, where several species of the *Alpina* are very much at home, but where none but the most ubiquitous of the *Subalpina, H. lingulatum* and perhaps *H. senescens,* are at all likely to occur.

The only two members of the *Alpina* group that extend to the south of the Scottish border are *H. holosericeum,* a species which is so widespread in the Scottish mountains that I have omitted to mention it above, and *H. subgracilentipes. H. holosericeum,* incidentally, has often been mistaken for *H. alpinum* itself and erroneously recorded as such.

Plate IX. Northern rock-cress, *Cardaminopsis petraea*. *a*. Small-flowered form. Side of River Avon near Tomintoul. (*M. C. F. Proctor*)

b. Large-flowered form. Ben Lui. (*R. M. Adam*)

M. C. F. Proctor

Plate X. An "alpine" hawkweed, *Hieracium holosericeum.* Snowdonia

It has several recorded stations in the English Lake District, and may yet, I suspect, be found in several more (see p. 111). Alone among the species of its section it is as likely to grow on an exposed summit plateau as on the cliffs below, and it is therefore worth looking for it anywhere above about 2,000 feet. *H. subgracilentipes,* apart from its localities on or near the Ben Nevis range, is known only in the immediate neighbourhood of Helvellyn, where it is by no means uncommon. *H. holosericeum* alone reaches Wales (Pl. X), while in Ireland no species of the section has ever been found.

The richest area that I know in Scotland for the section *Subalpina* lies in the range of hills running north and south to the east of Bridge of Orchy. Here, on a single short stretch of cliff, I have seen at one visit as many as six species of the section, *H. lingulatum, H. cremnanthes, H. sinuans, H. senescens, H. pseudanglicum,* and *H. gracilifolium.* Other places that are scarcely less rich, but where the variety of species is less concentrated, are the cliffs of Meall nan Tarmachan and Meall Ghaordie and—from all accounts, but I have never been there myself—Corrie Ardran on Ben More. Indeed, throughout the mountains of Perthshire several uncommon species, especially *H. sinuans* and *H. gracilifolium,* are locally abundant. The glens of Angus, too, are fairly rich, with *H. marshallii* tending to replace *H. sinuans.* Westwards, on the other hand, on the banks of the burns flowing into Glencoe, there is a widely different assortment. All the characteristic Perthshire species except *H. lingulatum* and *H. senescens* abruptly disappear from the picture, and are replaced this time, in places very abundantly, by *H. submurorum, H. callistophyllum* and *H. dasythrix.*

Of the four Scottish species which I have not so far mentioned, one, the Scandinavian *H. molybdochroum,* is known in Britain only on Cairntoul in the Cairngorms; one, *H. centripetale,* has its headquarters in the hills around Moffat, but has occasionally been found elsewhere in the Scottish hills; one, *H. hyparcticoides,* is confined to the banks of the Traligill (see p. 152) and possibly other burns in West Sutherland; and one, *H. isabellae,* is locally common in the glens around the Sow of Atholl. It might seem, therefore, that my earlier statement about the preference of the *Subalpina* for the mica-schist was an unwarrantable generalisation. It should, however, be remarked that none of the last three species named has very much in common with the rest of the *Subalpina,* and it may well be that they will ultimately have to be excluded from the section.

The British *Subalpina,* like the British *Alpina,* are nearly, but not quite, confined to the mountains of Scotland. The English Lake District possesses one speciality that is all its own, the very rare and puzzling *H. cumbriense;* but that is the only representative of the group that is found in either England or Wales. In Ireland, where the most interesting hawkweeds belong to the section *Tridentata,* the only species of the section *Subalpina* yet recorded (and that somewhat dubiously, since it is an abnormal form) is *H. senescens.*

These brief notes are necessarily of a very sketchy nature. For those who already have some acquaintance with the genus, the best I can hope is that some of the localities I have mentioned may be unfamiliar. For those for whom the genus *Hieracium* spells despair I have devised the key which will be found in Appendix II. Its aim, however, is very modest. For anybody who wishes to pursue the study of our mountain hawkweeds beyond the very simplest stage there remains only one possible course, the simultaneous study of Pugsley's monograph and of an adequate herbarium.

CHAPTER 7

VOLCANIC HILLS OF WESTERN SCOTLAND (J.R.)

WITH the obvious exception of the richest of the mica-schist mountains, which will be the subject of a later chapter, the most spectacular display of the British mountain flora is to be seen on the volcanic hills of the Western Highlands and islands. The Storr itself, towards the north end of the Isle of Skye, though perhaps the finest of all, is only one example of a type of hill by no means uncommon up the western side of Scotland. Other hills which are both geologically and floristically akin are Beinn na h-Uamha (to name but one of many) in Morvern, Ben Hiant in Ardnamurchan and Fionchra on Rhum. It happens, as a matter of fact, that none of these three should, strictly speaking, fall within the purview of this book at all. While the summit of the Storr does at least attain the respectable height of 2,360 feet, Beinn na h-Uamha is a mere 1,521, Ben Hiant 1,729 and Fionchra about 1,500. But since all three of them carry what is unquestionably a mountain flora, any sensible field-botanist, instead of adhering rigidly to an arbitrary definition, will rather feel grateful that, to see something of the Scottish mountain flora, he need not climb even as high as 1,500 feet; and since there can be few areas in Britain which have been so little botanised as these volcanic hills of the west, I should myself advise any botanist who wanted a rewarding holiday in the hills to turn his attention to this quarter. Here, certainly, he will see many of our loveliest mountain plants in splendid profusion; and here also, I suspect (and witness the recent discovery of *Koenigia islandica,* of which more anon), he stands as good a chance as anywhere in Britain of adding a new species to the list.

One of the most fascinating features of the particular hills which I have named is that, while for the most part they have a very similar flora, each of them has some special feature to distinguish it from the rest. Of the species which occur on all of them, and most abundantly of all on Beinn na h-Uamha, the most conspicuous are moss campion (*Silene acaulis*), purple saxifrage (*Saxifraga oppositifolia*), mossy saxifrage (*S. hypnoides*) and roseroot (*Sedum rosea*). These form, in every case, the luxuriant background against which the aristocrats of each hill are set. On Beinn na h-Uamha there are three species which vie for pride of place. Unfortunately my own explorations of the hill have been rather hurried, and I may therefore be wrong when I say that mountain avens (*Dryas octopetala*) is there local and scarce; and likewise alpine saxifrage (*Saxifraga nivalis*), of which I have seen, perhaps, some twenty plants in all on the hill, may well be more abundant on the stetches of cliff which I have not had time to explore thoroughly. But of holly fern (*Polystichum lonchitis*) there is no doubt at all: it is here as luxuriant and locally abundant as on any British hill I know. Incidentally, the bottom of the cliff in question can hardly be more than 1,300 feet above sea-level. Though further to the north both the saxifrage and the holly-fern come down even lower, I was still surprised to see them at so low an altitude so relatively far to the south.

On the cliff on the north side of Ben Hiant pride of place goes without question to northern rock-cress (*Cardaminopsis petraea*). Again I cannot speak with certainty, but I do not think that the rock-cress occurs at all on Beinn na h-Uamha. It is admittedly frequent on another rather similar hill in the same area, where it is accompanied in two or three places by the very rare Norwegian sandwort (*Arenaria norvegica*). But on Beinn na h-Uamha itself I have so far failed to find it; and if it were there at all, it would be there, I believe, as it is on both Ben Hiant and the Storr, in great quantity. There are in Britain two very distinct forms of this one species, the one small and compact, with thick and often reddish leaves and relatively tiny white flowers, the other much larger, indeed often almost straggling, with flowers nearly as large as those of the garden *Aubrietia* and sometimes faintly suffused with mauve. The former is the plant of Snowdonia, for instance, or the Cairngorms, while the classic locality for the latter, which has even been given the varietal name of var. *grandiflora,* is Ben Lui (Pl. IX, p. 128). On the Storr the plant falls squarely into the typical small form. But on Ben Hiant, as also on the Cuillins, there are numerous specimens which seem almost exactly intermediate between the two varieties. Nowhere else, admittedly, have I seen the extreme form which occurs abundantly on Ben Lui. But in view of these intermediate forms it seems very difficult to draw a clear line of demarcation between the type and the variety.

I have named Fionchra from among the many hills of Rhum because both in formation and in its flora it is most reminiscent of the Storr. The most notable plants of Fionchra, apart from alpine penny-cress (*Thlaspi alpestre*) which I shall reserve for discussion in a later chapter, are the alpine saxifrage, the two mountain grasses, *Poa glauca* (Pl. XVIIIa, p. 161) and *P. alpina*, and two species of eyebright. But even more interesting than Fionchra is another hill a few miles to the south called Ruinsival, on which Professor Heslop-Harrison and his party discovered the Norwegian sandwort. Ruinsival again, like the others, is of no great height, its summit being only 1,607 feet above the sea. It is a curious hill, the eastward ridge of which consists of a very bare stony wilderness broken by boulders and outcrops of rock. It is on these flat bare expanses that the sandwort grows, along with cushions of moss campion and mossy cyphel (*Cherleria sedoides*) and scattered plants of northern rock-cress. A more open community it would be hard to find; and that no doubt is what the sandwort likes. Anyhow it is in considerable plenty, and it has spread downwards from the ridge on to the shores of the fresh-water Loch Fiachanis some eight hundred feet below. Since it was discovered on Ruinsival it has been found again, this time by Messrs. E. C. Wallace and K. N. G. MacLeay, in yet another new locality, the hill in Morvern to which I referred earlier in this chapter. But even so, there are still only four localities in Britain, including one in the Shetlands, where the plant has yet been seen.

Almost exactly half way—and almost in a straight line too—between Ruinsival and the Storr stand the Cuillins; and though the Cuillins also are composed of an igneous rock, namely gabbro, it would be difficult to imagine two mountainous areas more different, both scenically and floristically, than those of the Cuillins and the Storr. The black bare cliffs of the Cuillins, with the long bare screes beneath, are to my mind seen at their best (from Arisaig on the mainland, for instance, or from the Isle of Canna) as a distant crested outline against the sky. It is of course a simple general rule that the mountains which are good for climbers are bad for botanists, and vice versa; and of all mountain ranges in Britain, the Cuillins would probably be ranked by the climbers who know them at the very top of the list. But for all that there is one plant of the Cuillins, alpine rock-cress (*Arabis alpina*), which is still unknown elsewhere in Britain. Originally found by a climber, in the person of H. C. Hart himself (see pp. 24-5), the alpine rock-cress is still apparently known to the climbing fraternity in a number of separate stations within the limits of the Cuillin range. But for myself—and the same is true, I suspect, of all other living botanists who have seen it—I know it in one spot only, a solitary rocky cleft which runs transversely across the face of a typically black and frowning cliff high above one of the two arms of Coire na Creiche. There is not a great deal of it in this locality, but what there is looks very flourishing. One large clump of it, certainly more than a foot in diameter, can be safely reached without climbing anything more severe than an apparently endless scree, while further up the cleft, which is overhung in its upper stages by a dripping projection of the cliff, there are some three or four other plants of an equal size. They grow in splendid isolation. Apart from a quantity of northern rock-cress scattered all over the same corrie, I saw no other plant of the slightest interest anywhere in the neighbourhood. Since botanists will doubtless continue to visit the Cuillins for the *Arabis*, it is highly desirable both that they should find it in another accessible locality and that they should find something else of interest with it to repay a stiff and gloomy walk.

If the Cuillins are the most hostile of British mountains, the Storr and the Quirang, for all their unearthliness, are among the most friendly; and nothing in Britain can surpass the ridge walk between the two, with its views of the mountains of Torridon to the east and those of Harris to the north-west. The best way to climb the Storr itself is up the north-east shoulder from the road, round the top of the great east-facing cliff, and up a gap in the smaller cliff which faces north on to the flat and boggy summit plateau. Every cliff, rock and scree on the latter half of this short climb has a rich and varied mountain flora. Northern rock-cress, moss campion, mossy cyphel, the saxifrages *S. oppositifolia* and *S. hypnoides*, rose-root, the grass *Poa glauca*, and green spleenwort (*Asplenium viride*) are all abundant; there is a liberal and unusual assortment of the three critical genera, lady's-mantle (*Alchemilla*), hawkweed (*Hieracium*) and eyebright (*Euphrasia*); globe-flower (*Trollius europaeus*), alpine saxifrage, alpine saw-wort (*Saussurea alpina*) and holly fern are locally more or less frequent; at the higher levels *Poa alpina* occurs, if in small quantity; and the summit plateau itself supports an abundance of three-flowered rush

(*Juncus triglumis*). Even without *Koenigia islandica* there can be very few hills in Britain of so lowly a height that can boast so rich a mountain flora (see Pl. XI, p. 144).

The story of the discovery of *Koenigia* in Britain is so unusual as to be worth telling in full. It was first found, on the summit plateau of the Storr itself, in the summer of 1934. The interests of its discoverers, however, were ecological rather than taxonomic, and the specimens of *Koenigia* which they collected were erroneously labelled *Peplis portula*. Under that name, with a quantity of other material, they found their way to the Herbarium at Kew, where they lurked undetected in the *Peplis* folder for more than fifteen years. It was late in 1950 that first Mr. B. L. Burtt in the *Kew Bulletin*, and then the *Times* itself, published the addition of *Koenigia* to the British list. With its circumpolar distribution, it was already known to extend far southwards, and even into the Southern Hemisphere, in certain of the loftiest mountain ranges; but until the winter of 1950 its nearest recorded station to the British Isles was in Iceland. It was in fact so interesting an addition to our flora that any botanist who had the opportunity was pretty certain to go during 1951 and try to rediscover it. It was duly rediscovered, in great quantity, not only in numerous localities on the Storr but also some six miles to the north along the same range. The fact that it had for so long eluded detection is an eloquent proof of how relatively neglected are these igneous hills of the west. It is probably a glacial relict species, as the recent discovery of its pollen in the Late-glacial peat-beds of Southern Scotland has conveniently confirmed (see p. 95 and Pl. XIIa, p. 145 (x3)).

In the only area in Britain where *Koenigia* is yet known it seems to be a very accommodating plant. The only condition that it apparently demands is an absence of competition. I have not seen it growing actually on the living rock; but wherever else on the Storr, from about 1,300 feet upwards, there is a really open community, there the *Koenigia* is as likely as not to be. On the long shoots of reddish stony soil, which are a prominent feature of the whole range, it grows in local abundance; and though it is here usually very small—not more than an inch at most in height or diameter—there are some places at the foot of these shoots, places where winter runnels fan out over semicircular patches of very fine earthy gravel, in which *Koenigia*, often accompanied here by the miniature cushions of awl-leaved pearlwort (*Sagina subulata*), may attain a respectable diameter of three inches or more. On the summit plateau, and likewise on the top of the ridge around the Bealach Amadal, it again seems to confine itself to bare, or nearly bare, stony patches. On the Storr itself its most constant companions in such situations are the three-flowered rush and the sedge *Carex demissa*. But away to the north, where the flat stony patches are for some reason noticeably drier, there are some places where *Koenigia*, though itself astonishingly abundant, is accompanied by nothing else except only the little moss *Oligotrichum hercynicum*. A typical stony patch in this area supports the following species :—

Agrostis canina	(bent)
Cerastium vulgatum	mouse-ear chickweed
Cherleria sedoides	mossy cyphel
Euphrasia sp.	(eyebright)
Festuca ovina	(fescue)
Galium hercynicum	heath bedstraw
Gnaphalium supinum	least cudweed
Koenigia islandica	
Sagina saginoides	alpine pearlwort
Thymus drucei	wild thyme
Viola riviniana	dog violet

There is nothing very remarkable about this list except the *Koenigia*; apart from that, such a community could be found on a great many of the Scottish hills. Since there are several other areas in the west of Scotland which are both geologically and floristically similar to the Storr ridge, and since they, even more than the Storr itself, have hitherto been largely neglected by botanists, there seems to be no good reason why the known range of *Koenigia* in the British Isles should not sooner or later be considerably extended.

This discussion of *Koenigia* inevitably brings to mind the two other arctic plants which have even more recently been added to the British list, *Diapensia lapponica* and *Artemisia norvegica*. Since neither of the two grows on an igneous mountain, they do not really belong to this chapter at all. But since also, by our division of the Highlands of Scotland into typical districts, there is really no chapter in the book into which they can be properly fitted, this is probably the best place to say a little about them. As members of the British flora they have at least one accidental characteristic in common, both with one another and with *Koenigia,* that they were first discovered in their new localities, not by a professed botanist in search of rare species, but by people whose primary motive for climbing the hills in question was, if not quite different from that, at least less ambitious. Indeed, the mountains that support these two plants are both for the most part so botanically unpromising that no botanist in quest of a new British plant would have been likely to give them more than one cursory look; which seems to suggest that on some of the less obviously inviting of Scottish mountains other arctic species may even now be awaiting discovery.

The home of *Diapensia lapponica* is a single hill in the neighbourhood of Glenfinnan, to the west of Fort William. There it was first found, early in July, 1951, by Mr. C. F. Tebbutt, a naturalist whose interests are primarily ornithological. It was surprisingly still in flower, and specimens of it were sent to Kew, to Edinburgh and to Aberdeen. It occupies a mere acre or so of a rocky crest of hill at a height of a little over 2,500 feet (Pl. XIII, p. 148). The hill as a whole is composed, like most of the vast and little known tract of country to the north of Glenfinnan, of one of the more barren forms of schist, and apart from the *Diapensia* the best that it can produce in the way of a mountain flora consists of moss campion, two "alpine" hawkweeds (*Hieracium holosericeum* and *H. gracilentum*), bearberry (*Arctostaphylos uva-ursi*), mountain azalea (*Loiseleuria procumbens*) and least willow (*Salix herbacea*). Devil's-bit scabious (*Succisa pratensis*), golden-rod (*Solidago virgaurea*) and (slightly more promising, perhaps) cat's-foot (*Antennaria dioica*) are all frequent. It is in fact a typical flora of a dull ridge. But in the middle of this dull ridge the nature of the rock suddenly and obviously changes, and for fifty yards or so there is a stretch of curiously convoluted outcrops, interspersed with stony patches reminiscent at first sight of the sugar limestone of Widdybank Fell in Teesdale. On this very limited area *Diapensia* abounds, great spreading patches of it, two feet and more across and sometimes dying away in the middle like old rock-roses in a garden, alternating with young and compact cushions. I have not yet succeeded in seeing it in flower; in 1952, when I went to pay it my respects, it was well over by June 20th, while in 1953, a friend found it in full flower a whole month earlier than that. No doubt its flowering season varies widely from year to year with the duration of the covering of snow on the hill. But when it is in full flower it must present, with its large creamy-white flowers rising in thousands from the dense olive-green rosettes, a sight that few mountain plants in Britain can surpass (Pl. 12, p. 125).

The discovery of *Artemisia norvegica* in Britain, though its first finder had seen it in its new locality two years earlier without realising its significance, was only confirmed as recently as August, 1952. It too grows along the top of a ridge, this time in Wester Ross, and at very nearly the same altitude as *Diapensia*. It is, I suppose, the most notable of the three recent additions to our mountain flora; for whereas the other two do at least occur in Iceland as well as in Scandinavia, the *Artemisia* had hitherto been regarded as confined to the district around the Dovrefjeld in Norway and to the Northern Urals. I do not think I can do better, by way of describing the new locality and the plant's behaviour there, than to quote from the original description of it (published in Mr. R. A. Blakelock's excellent article in the *Kew Bulletin,* 1953) which was based on the notes of the original finder himself, Sir Christopher Cox.

"The species was growing on the spur of a mountain between 2,350 and 2,450 feet (possibly as low as 2,250 feet in one place). The spur forms an undulating plateau rather less than half-a-mile long and not much more than 100 yards wide at its narrowest but some 350 yards or more wide at its broadest point, where the slight tilt from N.E. to S.W. which characterises the whole spur becomes much more pronounced and the plateau forms a kind of sideways-slipping basin above what seems to be a break in the line of crags below. More than half the colonies observed were in this basin. Twenty-four colonies in all were counted, each say 5 to 10 yards square; within each colony examined the plant was growing fairly plentifully in up to half a dozen small areas, each such area containing one to (usually) several clumps or single plants or, more commonly, single plants and clumps together. The largest contiguous clump measured was 14 cm. by 4–6.5 cm. On the assumption that contiguous but apparently separate plants in small clumps were in fact separate, the collector considers that, on a very rough estimate, he saw between 500 and 1,000 plants, and he has little doubt that, as it was only possible to examine parts of the spur in the limited time available, a number of colonies exist in addition to those which he saw.

"There was a marked lack of flowering stems—several colonies and many clumps had none. The type of flowering stem seen most often was small and seemingly shrunken or withered, giving the appearance of a withered bud. Of fully grown flowering stems the more usual type was the solitary head; the double-headed specimens, though more than one was picked for the vasculum, were much rarer. The most notable single cluster of flowering specimens comprised four contiguous plants, each with single-headed flowering stem, side by side, the heads all turning to the S.W., the whole forming a clump about 9 cm. wide, and about 20 cm. away two contiguous plants with double-headed flowering stems.

"*Artemisia norvegica* is thought to have been about the third commonest plant of the type of habitat where it occurred, the other two being *Alchemilla alpina* and *Antennaria dioica,* not taking into account the abundant small tufts of grass. *Gnaphalium supinum* was not infrequent in the vicinity.

"A list of the associated plants is:—

Cherleria sedoides	*Solidago virgaurea* var. *cambrica*
Alchemilla alpina	*Vaccinium myrtillus*
Galium hercynicum	*Thymus drucei*
Succisa pratensis	*Salix herbacea*
Gnaphalium supinum	*Carex* sp.
Antennaria dioica	*Deschampsia flexuosa*

The surface where the plants occur is a "clitter" of low flat boulders and stone with patches of bare gravel or gravelly soil in between (the patches occupying about as much space as the large stones), and occasionally outcrops of the living sandstone piled in characteristic layers. The habitat is exposed and normally dry. . . .

"Rock samples were kindly examined by the Geological Survey and Museum, S. Kensington, who send the following report. The samples 'are of felspathic grit from the Torridonian sandstone formation. Petrographically they belong to the type of grit known as *arkose,* that is, a felspathic grit derived by the denudation of granite or gneiss with little intervening decomposition. The rock is thus composed essentially of felspar and quartz' . . .". Finally Mr. Blakelock adds a soil analysis by the Macaulay Institute for Soil Research: " . . . The soil is markedly acid (pH 4.8) and very low in calcium. It is also very low in phosphate. The potassium and magnesium contents are relatively much higher, being comparable with the values encountered in many agricultural soils."

This long account does so admirably for *Artemisia norvegica* what needs to be done for many another of the rarer mountain plants that it would be an impertinence to add much to it. Beyond remarking that the *Artemisia* seemed to Max Walters and myself to be even more abundant than Sir Christopher Cox had suggested, I shall merely append, by way of confirmation and expansion of the original list, another list of associated plants made in the same locality by Messrs. J. E. Lousley and D. McClintock. I am very grateful to Mr. Lousley for permission to reproduce it and his photograph (Pl. XIIb, p. 145).

Alchemilla alpina	alpine lady's mantle	f
Antennaria dioica	cat's-foot	o
Cherleria sedoides	mossy cyphel	o
Deschampsia flexuosa	wavy hair-grass	o
Festuca vivipara	viviparous fescue	f
Galium hercynicum	heath bedstraw	o
Gnaphalium supinum	least cudweed	r
Juncus trifidus	three-leaved rush	r
Luzula multiflora	(wood-rush)	r
Polygonum viviparum	viviparous bistort	v.r
Salix herbacea	least willow	f
Solidago virgaurea	golden-rod	r
Thymus drucei	wild thyme	f
Vaccinium vitis-idaea	cowberry	v.r

Finally, as a postscript to *Artemisia,* it is interesting to add that the Swedish botanist Hulten (1954) has now distinguished the British plant as an endemic variety *scotica,* differing in habit and leaf-shape from the Norwegian type.

CHAPTER 8

LIMESTONES OF NORTHERN ENGLAND AND NORTH-WESTERN SCOTLAND

(*a*) Northern England (J.R.)

The carboniferous limestones of Northern England, which extend from the north shore of Morecambe Bay as far as Upper Teesdale and beyond, have already been described in two chapters of Mr. J. E. Lousley's book in this series, *Wild Flowers of Chalk and Limestone*. Since, however, Mr. Lousley deliberately omits the alpine plants of the higher summits on the ground that they belong to this rather than to his own book, some mention must be made at this point of a number of species especially characteristic of this particular region. I shall restrict my remarks to two small portions of this one large area, namely the neighbourhood of Ingleborough in the West Riding of Yorkshire and the hills on either side of the River Tees above Middleton in Teesdale; and I shall concentrate particularly on those species which, if not actually confined to these two areas, are at least more easily found here than anywhere else in Britain.

The most spectacular element in the flora of the upper parts of Ingleborough, as also of Penyghent to the west, is undoubtedly the purple saxifrage (*S. oppositifolia*), which makes as vivid a show here as anywhere in the British Isles. Near the summit of Ingleborough it is accompanied on the cliffs by the very rare non-viviparous form of the alpine meadow-grass (*Poa alpina*), a form which is at present known elsewhere in Britain only on the cliffs and gullies of Caenlochan Glen in Angus. Other mountain species on Ingleborough (see Ch. I, p. 9) are alpine penny-cress (*Thlaspi alpestre*), hoary whitlow-grass (*Draba incana*), alpine cinquefoil (*Potentilla crantzii*) and alpine bistort (*Polygonum viviparum*). But the chief distinction of Ingleborough lies elsewhere—in its assortment of lady's mantles (*Alchemilla* spp.) and its possession of a sandwort unknown anywhere else in Britain, *Arenaria gothica*.

Discussion of the lady's mantles is best reserved for a special note on the subject: suffice it for the moment to say, first, that *Alchemilla minor,* with its curiously discontinuous distribution in Britain, is more abundant around Ingleborough, on grassy tracks and banks, than in any other of its few British stations; and second, that *Alchemilla minima,* recently described by Max Walters, is not at present known anywhere else in the world. The same cannot be said of the sandwort, which is known outside Britain in both Scandinavia and Switzerland (see p. 64). Nor indeed is it yet agreed that *Arenaria gothica* is specifically distinct: it and *A. norvegica* are still regarded by some botanists as subspecies of the one variable species *A. ciliata*, the Irish form of which has been named subspecies *hibernica.* There is moreover this much to be said for those who regard the three as subspecies rather than species, that a comparison of the distribution in Britain of the three together with that of *Alchemilla minor* shows the most remarkable coincidence. In one of the two Scottish localities for *Alchemilla minor*—and the two are not many miles apart—grows *Arenaria norvegica*; on Ingleborough *Alchemilla minor* and *Arenaria gothica* grow together, each of them restricted in England to that immediate neighbourhood; and in Ireland *Alchemilla minor* and *Arenaria ciliata* are both alike confined to the limestone hills to the west and north-west of Sligo. The three sandworts would seem therefore to have very similar requirements, each of them in fact choosing damp stony slopes, sometimes liable to winter inundation, on the sides of limestone hills or ridges. On Ingleborough, or at least on its east-facing side, *Arenaria gothica* is said to be thinly scattered over a tolerably wide area; but for myself I know it only on one small and very damp stony patch near the foot of the mountain, where the associated species include vernal sandwort (*Minuartia verna*, Pl. VI, p. 97 (×1½)) and hairy stonecrop (*Sedum villosum*).

The flora of Teesdale is perhaps better known—and by no means only to professed botanists—than that of any other mountainous area in Britain. The local abundance of such species as shrubby cinquefoil (*Potentilla fruticosa*), bird's-eye primrose (*Primula farinosa*) or spring gentian (*Gentiana verna*) gives to the valley of the Tees a glamour that can be rivalled in the British Isles only by the Burren in County Clare. None of these three, however, is in Teesdale a primarily montane plant, each of them preferring the banks of the river and its tributaries to the cliffs or summits above. More definitely montane in their choice of habitat are the two species which (as is well known from the Floras) are confined to the sugar limestone of Widdybank Fell, namely Teesdale violet (*Viola rupestris*) and bog sandwort (*Minuartia stricta*, Pl. V, p. 96 (×1½)). Both, I suspect, are commoner on Widdybank than is often supposed, but they seem to have different requirements. Whereas the sandwort chooses, like its relative on Ingleborough, bare stony patches which are perpetually very damp indeed, the violet seems to prefer the ribs of living rock which just project on the slopes from the close turf. I still have a feeling myself that either species, and especially the violet, is liable to be found at any moment in the neighbourhood of Inchnadamph, and I even have my eye on the particular little valley in which Max Walters and I lately found *Alchemilla minor* (see p. 151). We were too late in the year on that visit to hope for any recognisable traces of the sandwort, while a violet which we found in some quantity is still under critical observation; but another visit in May or early June might yet, I believe, prove this particular hunch of mine to be well founded.

Other mountain plants growing within a mile or so of the Tees include hoary whitlow-grass (*Draba incana*) in the very small form once called var. *nana*, mountain avens (*Dryas octopetala*), alpine bartsia (*B. alpina*), two rushes (*Juncus alpinus* and *J. triglumis*), the sedges *Kobresia simpliciuscula* and *Carex capillaris*, and the holly fern (*Polystichum lonchitis*). This last is now, thanks wholly to collectors, exceedingly scarce on the cliff where it has for so long been known,—a fine plant that I saw there not long before the war had disappeared when I last went to look for it,—while another and even rarer fern which used to inhabit the same cliff, *Woodsia ilvensis*, has not to my knowledge been seen for so many years that I fear it must probably be written off as extinct in the locality—another victim of the avidity of collectors. This particular station for the ferns, which is more accessible than most, suffered the misfortune of too much publicity. The fate of the *Woodsia* is a sad warning against precision in the publication of localities for the rarest plants.

Further afield from the Tees there are three other plants of which some mention would have to be made in even the most summary account. Of the beautiful purple-awned grass *Phleum commutatum*, I need only say that it still lingers on in the hills not far from the Tees—the only British localities outside Scotland where it has been seen in recent years. Likewise the alpine forget-me-not (*Myosotis alpestris*) calls for little comment. On at least one stony slope high up in the hills it is,

I believe, still abundant; but I must admit that the length of the walk involved, and through the deepest heather too, has hitherto deterred me from going to see for myself. But as for the third of the trio, yellow marsh saxifrage (*S. hirculus*, Pl. XXIV, p. 189), that needs somewhat fuller treatment; for the literature on its distribution in Britain, and on the type of habitat it selects, is liable to create a false impression.

The yellow marsh saxifrage (see Appendix III, map 8a) is still to be found in as many as nine vice-counties in Northern England or Scotland and six in Ireland. But, for all that, I should be surprised to hear of any living botanist who has seen it in more than one of these areas. Again, it would be easy to gather from some of the Floras that the place to look for it is in the wettest of *Sphagnum* bogs; and it may well be, to judge from Praeger's *The Botanist in Ireland*, that its Irish localities, on the Garron Plateau, for instance, or in County Mayo, are usually of a rather different nature from its English (see Ch. 10). But certainly, if you want to find it on the Yorkshire moors to the south of Teesdale, you had best forget about *Sphagnum* bogs and search for it instead on rushy hillsides with patches of damp grass between the clumps of rushes. In such places, especially on the moors between Barnard Castle and Brough, it is not, I believe, so desperately uncommon; at all events each of the three parties that I know to have lately gone in search of it has duly found it, and each in quite different parts of the moor. Its associates where I saw it included such plants of the grassland as meadow buttercup (*Ranunculus acris*) and self-heal (*Prunella vulgaris*); others such as lesser spearwort (*Ranunculus flammula* ssp. *flammula*) and the heath rush (*Juncus squarrosus*) which indicated a certain degree of moisture; but only one, hairy stonecrop (*Sedum villosum*)—incidentally a plant with the agreeable habit of growing in the most distinguished company—which might suggest that the place was in any way unusual. The saxifrage itself is said to be a great delicacy for sheep. It may have been because the heavy snowfalls of the previous winter had decimated the flocks on the Yorkshire moors that, in the summer when I saw it, it was flowering freely enough.

Finally, no account of the mountain flora of the Yorkshire limestone should omit the mountain pansy (*Viola lutea*), which is so abundant in the rough bent-fescue grassland both in the Malham area and in Upper Teesdale. This decorative perennial pansy was known to all the early botanists: it figures in Gerard's Herball (1597) as 'the yellow violet . . .

J. E. Raven

Plate XI. The Storr, habitat of *Koenigia islandica*

Plate XII a. *Koenigia islandica* *J. E. Raven*

b. *Artemisia norvegica* var. *scotica* *J. E. Lousley*

upon the hills in Lancashire neere Latham', and was familiar to John Ray 'in many places among the mountains . . .' (Ray, J. 1670). In very recent years also a good deal of study has been devoted to the mountain pansy, including a series of papers on the cytogenetics of a naturally-occurring hybrid population involving the common wild pansy *Viola tricolor* (Fothergill, P. G., 1941, and earlier papers quoted there), and an excellent autecological account in the Biological Flora series (Balme, O. E., 1954). We therefore know more about the mountain pansy than about many of our mountain plants, including a good deal about its soil preferences, which go far to explain its remarkably patchy local distribution.

Strictly speaking, the majority of the localities for the pansy would not come within the scope of this book, for they are well below 2,000 feet. Nowhere is the plant more abundant, for example, than on the Derbyshire limestone (where most of Miss Balme's detailed study was centred); its altitudinal range in West Yorkshire is given by Lees (1888) as 450 to 1,800 feet, and in Ireland it ranges from sea-level to 1,250 feet only. In the Scottish Highlands, however, it qualifies without any doubt, occurring for instance on the Lawers range up to 3,450 feet (White, 1898) on rock-ledges with a southern exposure. The Lawers plant is a magnificent pansy too; the flower is often quite remarkably large, well over an inch long, and usually deep purple. The distribution of this purple-flowered form (variety *amoena*) seems to have a clear geographical pattern: much of the Scottish material is purple, whilst the yellow form definitely predominates in Derbyshire, where small patches of var. *amoena* occur within the typical yellow-flowered populations. Actually, however, the two colour-forms are not clearly distinct, for populations also occur with flowers with the lateral petals purple and the rest yellow. Even in the fully purple flower the central 'eye' is always yellow.

The mountain pansy occurs in almost every vice-county in Great Britain north and west of a line from the Severn to the Humber, and in six Irish vice-counties also (see map in Balme, 1954). This distribution is clearly relatable to a preference for upland areas where there is relatively high rainfall and cool summer temperature. On the Continent, it occurs in France, Belgium, Germany, Switzerland and Austria, and further east as a distinct subspecies. It is significantly absent from Scandinavia, and could be included in the alpine element of our mountain flora.

Although the mountain pansy is abundant on the Yorkshire and Derbyshire limestones, and rarely occurs on really poor, siliceous soils, Miss Balme's work showed that it is not really a calcicolous species. It proved, in fact, to be absent from highly calcareous soils, such as those of unstable limestone 'scree' slopes, and most abundant on reasonably stable mineral soils where some leaching had removed the free calcium carbonate from the surface layers of the soil. Miss Balme was further able to show that the inability of the plant to grow properly on highly calcareous soils was probably due in part to its being unable under these conditions to take in sufficient potash and phosphate for healthy growth. The patterns shown by the mountain pansy in limestone areas can largely be related to these rather strict soil preferences.

There is no difficulty in recognising the mountain pansy in most British stations by its large long-spurred flowers borne on long stalks, and its slender creeping rhizome. Fothergill's work, however, has shown that where the mountain pansy meets forms of the common wild pansy, *Viola tricolor*, a variable hybrid population can arise in which the individuals can show many different combinations of the characters of the parental species. Such hybrid populations exhibit a very complex cytogenetic make-up, which Fothergill was able to analyse to some extent. The chromosome numbers of the two species are $2n = 48$ (*V. lutea*) and $2n = 26$ (*V. tricolor*), and the hybrids show numbers ranging from 24 to 58. There seems little doubt that extensive hybridisation in the wild pansies is responsible for a great deal of the difficulty of classifying the group, and the mountain pansy is relatively distinct, probably only because it 'keeps itself to itself' by means of its strict climate and soil preferences. It is interesting that the perennial wild pansy which occurs abundantly on many western sand-dunes, which has been variously called a species, *Viola curtisii*, or a subspecies of *V. tricolor*, resembles the mountain pansy in flower colour and in perennial habit, but is not normally confused with it because the two do not occur together. In the west of Ireland, however, where, as we shall see again in Chapter 10, mountain, coastal and lowland plants can all freely intermingle, the two pansies do meet, and apparently give the expected intermediates, as Syme pointed out many years ago. Indeed, a century of careful endeavour to classify wild pansies has got us no further than Syme; the pansies are still (Syme, 1863, Vol. 2, 28) 'a good instance of the convenience of the employment of subspecific groups', and it is still an open question whether we would not do better to include, as Syme did, the mountain pansy as the fourth subspecies of the aggregate wild pansy, *V. tricolor*.

Note on Mountain Lady's Mantles (Alchemilla). (M.W.)

Most upland areas in the British Isles carry their populations of lady's mantles, though if one is interested in seeing profusion and variety, one had better avoid North Wales, the Irish mountains, and the Cairngorms. Indeed, as we have indicated, some of the well-worn botanical paths—Ingleborough, Teesdale, and the Lawers range—should be trodden by the student anxious to learn his way about this rather perplexing group. I have already shown (Ch. 4, p. 76) something of the interest and significance of the mountain *Alchemillas* in the British mountain flora as a whole; it would now perhaps be worth while to give general hints and suggestions to any who may be inspired to use the short key (Appendix II) to identify the *Alchemillas* they may meet on British mountains.

The neat little alpine lady's mantle (*A. alpina*) we have already mentioned several times; it is locally abundant in the Scottish mountains, particularly in the west, and in the Lake District, but unaccountably absent from Snowdonia and extraordinarily rare in Ireland. All British plants are, so far as we know, essentially similar, and belong to a widespread arctic-alpine species. Its sturdier distinguished relative, *A. conjuncta* (Pl. XVI, p. 157)—which is often grown as 'alpine lady's mantle' in gardens—is a handsome plant with its leaflets dark green above and silvery beneath, as in *A. alpina*, but the leaflets are obviously joined for a ¼ inch or so at the base. There are two places only in the whole of the British Isles where this lady's mantle is generally held to be native. One will be described later—Glen Clova (Ch. 9, p. 179); the other is Glen Sannox, in the Isle of Arran. *Alchemilla conjuncta* has been recorded from a good many other places, including the Lake District mountains; but all these records are thought to refer to garden escapes or to be due to some confusion of identity. It is a remarkable fact that a plant which grows so happily both vegetatively and from seed in most parts of the British Isles should be so rare as a native plant—so remarkable that one cannot entirely dismiss the possibility that even in Clova and Arran it was originally introduced.

These two plants are very easily recognised both from all other mountain plants (by their silvery-backed leaflets) and from each other.

When we come to the so-called 'common' lady's mantles, however, there is more difficulty. Eleven different species are described in the British Flora of Clapham, Tutin and Warburg, and although it is not difficult to name well-grown specimens which are flowering freely, it is often not possible, particularly in grazed mountain pasture or late snow patches, to find well-grown complete specimens. Fortunately four species only of this group ascend commonly above 2,000 feet on British mountains, namely *A. filicaulis*, *A. glabra*, *A. glomerulans*, and *A. wichurae*; so that a simple key can be constructed which will normally enable one to identify a mountain *Alchemilla* from a single well-grown leaf (see Appendix II). A word of warning should however be given concerning the use of the key on the Breadalbane range. There seems to be a rather mysterious *Alchemilla* on these mountains which has leaf-characters inconveniently halfway between *A. glabra* and *A. wichurae*; its status and identity are not yet determined, but its presence has the effect of blurring the distinction between these two species, particularly in the Lawers area. In other parts of the British mountains there is little or no difficulty in separating the much rarer *A. wichurae* from the often abundant *A. glabra*.

None of the other British *Alchemillas* are really mountain plants; they are not to be found, as are those previously mentioned, in the rich herb vegetation in steep mountain gullies or on high wet rock faces, but are rather plants of hilly regions. From the point of view of choice of habitat they can be divided into two types, those characteristic of grazed pasture, and those characteristic of hay meadows, roadsides or, more naturally, river-margins and open marshy woodland. *Alchemilla minor*, already mentioned, is a good example of the first class; it is a neat, hairy, dwarf plant, well adapted for growing in the short open turf of the Ingleborough limestone pavement, where it occurs abundantly between about 1,000 and 1,500 feet. A much more extreme 'dwarf' is the tiny blue-green *A. minima*; this seems to be closely related to the most widespread lowland species, *A. vestita*, and may well have originated by mutation from this species under conditions of close grazing by sheep, where dwarfness is a great advantage.

The other, taller and coarser group of roadside or hay meadow species includes the widespread *A. xanthochlora*, which is the commonest *Alchemilla* in the many parts of Northern England and Lowland Scotland where the soil is acid, and *A. glabra*, which we have already

R. M. Adam

Plate XIII. Habitat of *Diapensia lapponica.* Scotland

M. C. F. Proctor

mentioned as occurring high on mountains. In hilly regions, particularly where there is some limestone, *A. glabra* is usually as abundant on the valley roadsides as it is on the mountains. In addition to these common roadside and meadow species, however, Upper Teesdale and Weardale possess a unique and remarkable set of three species, common in the mountains of Central Europe and indeed over much of Scandinavia also, but otherwise absent from the whole of the British Isles. One of these, *A. monticola*, which is locally abundant along the roadside and in hay meadows in Upper Teesdale, is the common lady's mantle of the Swiss alpine valleys. It is a most remarkable fact of plant geography that these common Central European *Alchemillas* should appear in some abundance locally in and around Teesdale, thus putting the finishing touches on the striking resemblance of the rich Teesdale meadows to the sub-alpine meadows of Switzerland.

Teesdale is not, however, content with this distinction, but produces also all the other truly mountain British *Alchemillas*. The occurrence of *A. glomerulans* is the most remarkable; for whilst in the Scottish Highlands this lady's mantle is strictly a plant of the high cliffs, only rarely descending much below 2,000 feet and then washed down by rivers, in Upper Teesdale it occurs in several places sitting prosaically with all the other *Alchemillas* on the grass verge by the main road (Pl. XV, p. 156). Although there is really no difference in form to be detected between the Teesdale and the Scottish Highland plants, one feels that there must be some internal difference to account for the much more accommodating behaviour of the Teesdale population. In cultivation in Cambridge the Teesdale plants certainly grow rather more vigorously than their Highland Scots relatives.

(*b*) North-Western Scotland (J.R.)

If there is any part of Scotland, indeed of Britain as a whole, more hauntingly lovely than the Northern tip of Skye, there is no doubt in my own mind where it lies. It is that rectangle of mountain, moorland and loch, which is bounded on the north by Kylesku and Clashnessie, on the south by Elphin and Achiltibuie. From an intricate pattern of lochs and lily-lochans, of rocky hillocks and wooded glens, there arise within this area seven of the shapeliest of all mountains. To the south, in a triangle, stand Cul Mor, Cul Beag and Stack Polly, each, despite their proximity, with an isolated individuality of its own.

In the middle, in contrast one to the other, rise the sharp cone of Canisp and (seen from Lochinver at least) the blunt parabola of Suilven. To the north, all by itself, is Quinag; and Quinag, unlike the others, contrives to combine four or five distinct summits into a single whole. For those who like to be able to view their mountains, rather than merely sensing that they are there, this is the perfect country. Each one of these seven can not only be seen as an individual mountain, it can be seen, and without even leaving the road, from every point of the compass. And if there are some people for whose taste the shapes of these hills are too dramatic and grotesque, yet it is surely better that a hill should err in that direction than that it should have no shape at all.

If a botanist were to start at Ledmore Lodge and drive around the road which encircles this area, he would probably notice a number of spots by the roadside where the rock became whiter, the grass greener and the meadows more full of flowers. There are three such places in particular where the presence of limestone is unmistakable. One is near Stoer, where the roadside banks abound with globe-flower (*Trollius europaeus*), bitter vetch (*Vicia orobus*) and burnet rose (*Rosa spinosissima*). Another is at Knockan and Elphin, where the long cliff to the south-east of the road is a well-known locality for several of the plants especially characteristic of this area. Third and last is Inchnadamph itself; and if anybody wishes to see in its most obvious form the difference between the vegetation of an acid and of a basic soil, then he can do no better than to walk from Inchnadamph to the junction of the Kylesku with the Lochinver road, and there, facing towards Kylesku, consider the country on either side of him. On his right is white rock and green grass, on his left grey rock, heather and peat. And if he wants to see the rare plants of the district, of which there are many, he will thereafter cling firmly to the former.

There are few botanical centres in Britain more rewarding than Inchnadamph itself, nor any better suited to the indolent. Within a hundred yards of the hotel, on the banks of the Traligill Burn, grow fine clumps of the rare grass *Agropyron donianum*, not to mention an abundance of *Trollius*. Within a quarter of a mile, on the long limestone cliff running southwards, not only purple saxifrage (*S. oppositifolia*) and the *Agropyron* but dark red helleborine (*Epipactis atrorubens*) and the sedge *Carex rupestris* are in unusual plenty. Nearby, in a locality which it is perhaps unwise to define too closely, grows

Norwegian sandwort (*Arenaria norvegica*). Barely half a mile to the north of the hotel, on the rabbit-grazed south-facing slopes of a shallow valley, is one of the only two known Scottish stations of *Alchemilla minor*; while on the flat bottom of the valley, which consists of very wet stony patches beside a stream, there grows, amidst a mass of yellow mountain saxifrage (*S. aizoides*), a little of the hair sedge (*Carex capillaris*, Pl. VIIb, p. 112). Add to this that mountain avens (*Dryas octopetala*) is everywhere abundant, and the attractions of the place should be obvious.

Agropyron donianum, despite its local abundance, has been known to grow at Inchnadamph only since 1951. Until then its only known stations were certain alpine cliffs within a few miles of the summit of Ben Lawers. It had actually been collected in the Inchnadamph neighbourhood by the Rev. E. S. Marshall as long ago as September 12th, 1887; but Marshall himself, taking it for nothing but an abnormal form of bearded couch-grass (*A. caninum*) and labelling it accordingly, had evidently put it away in his extensive herbarium and forgotten it. Marshall's herbarium is now in the Botany School at Cambridge, where it is consulted pretty regularly for its excellent collection of the mountain plants of Scotland, especially the hawkweeds; and it was natural enough that Max Walters and I should have turned to it, as we did in 1951, for light on the question of what the nineteenth-century botanists meant by *Agropyron caninum* var. *alpinum*. So it was that the presence of *A. donianum* at Inchnadamph was first detected. Unfortunately, besides misnaming his plant, Marshall had also described its locality on the label only in the very vague terms "Near Inchnadamph"; so it looked for a while as if the task of rediscovery might be a long one. Marshall, however, had a convenient habit of writing up in the *Journal of Botany* an account of the most notable of his previous summer's activities; and though the account published in 1888 proved to contain no reference at all to any *Agropyron*, it did contain some comments on two other plants, the willow *Salix myrsinites* and a hybrid willow-herb, which he had collected during his stay at Inchnadamph. Further reference to his herbarium not only showed that these two plants were collected on the selfsame day as the *Agropyron*, but provided us also with two precise pinpoints on a map between which he must have walked on the day in question. It was highly satisfactory to discover six weeks later, when the opportunity came to test these herbarium researches in the field, that their conclusions were accurate enough. Marshall had walked from the hotel up the banks of the Traligill Burn, on which he had gathered *Salix myrsinites*, on to the slopes of Ben More Assynt, where he found his hybrid willow-herb. An exactly similar walk, seventy-four years later, revealed that the *Agropyron*, though frequent all down the lower reaches of the Traligill, is particularly so at precisely the spot where Marshall collected the willow (Pl. XVIIIb, p. 161).

The Traligill is one of those remarkable burns, not uncommon on the limestone, which at one point in their course suddenly disappear into the ground to leave their rocky beds for a while quite dry. The best walks around Inchnadamph, from the botanist's point of view, follow the footsteps of Marshall at least as far as the cave into which the river vanishes. Then, however, there are several tempting possibilities, and the only guiding principle is to go wherever the limestone comes up through the peat. For my own part I tend to make for a particular limestone ridge, which juts out for half a mile or so from the side of an otherwise acid hill; for this ridge is one of the centres of the mountain flora of the district. The broken cliff which forms its flank is interesting enough, with an abundance of moss campion (*Silene acaulis*), mountain avens, the saxifrages *S. aizoides* and *S. oppositifolia*, and an occasional vigorous plant, looking rather gross among its companions, of stone bramble (*Rubus saxatilis*). But the broad top is even richer. Not only is this the real home of the Norwegian sandwort, which grows here, just as on Rhum, chiefly on open patches of soil and stones between the outcrops of rock, but it supports too (even if in so ridiculously small a form, due to rabbit-grazing, that they are scarcely recognisable as such) occasional colonies of alpine cinquefoil (*Potentilla crantzii*) and alpine saw-wort (*Saussurea alpina*), as well as *Carex rupestris*, green spleenwort (*Asplenium viride*), and holly fern (*Polystichum lonchitis*). With the possible exception of the rocks near the mouth of the Naver, which I shall be describing soon, this is the finest example to be seen of the flora of the north-western limestones.

At the point where this limestone ridge runs into the side of the acid hill, there is a change in the vegetation quite as abrupt, if not so conspicuous, as that by the Kylesku road. Mountain avens and the rest suddenly give way to a flora consisting predominantly of heath, heather and bilberry. But even on this acid hill there are still interesting plants to be seen. Where the slope of the hill is drier and more stony, both black bearberry (*Arctous alpina*) and mountain azalea (*Loiseleuria procumbens*) spread their horizontal branches, the leaves of the former in the autumn splashing the grey stones with blood-red blotches; and where it is wetter, cloudberry (*Rubus chamaemorus*) and dwarf cornel (*Chamaepericlymenum suecicum*) take their place. With least willow (*Salix herbacea*) at the summit, this single hill can muster, for all its apparent barrenness, a quite notable collection of truly dwarf shrubs; and one of that collection, the black bearberry, which I have not had occasion to mention before, is a plant of sufficient interest to justify a short digression.

The first interesting fact about *Arctous alpina* is, as its name might seem to imply, its exceptionally northern distribution in the British Isles. Apart from a number of stations on or near Ben Nevis, notably the edges of the summit plateau of Ben Chlianaig near Roy Bridge, it seems (at least so far as recently confirmed records are concerned) to be entirely confined to the part of Scotland north of the Caledonian Canal (see Appendix III, Map 2a). Even within that area, though it is not especially uncommon, it seems to be unusually particular in its tastes. It grows nearly always (in my experience at least, which does not include Orkney or Shetland) within two or three hundred feet on either side of the 2,000 foot contour; and the sites it chooses, therefore, are normally not so much the summits of mountains, but rather, as on the hill near Inchnadamph, bare stony shoulders or ridges. Such at any rate are its localities on and near Ben Nevis, on the hills above Glen Affric, where it grows with its close relative *Arctostaphylos uva-ursi*, and on such a hill of the far North as, for instance, Ben Klibreck near Altnaharra. In one respect the Inchnadamph locality differs slightly from the normal; it is, as several others of its most northerly localities are, somewhat lower than the usual lowest limit of 1,700 to 1,800 feet. Otherwise it is as typical as it could be. Seeing, then, that such are its chosen localities, and that it does at least spread far enough to the south to be perfectly at home on Ben Nevis, it is a puzzling question why the Cairngorms, which afford acres of apparently ideal territory, should be entirely devoid of it. The map of its distribution in Britain provides in fact a striking contrast to that of *Carex rariflora* (Appendix III, map 3b).

Another remarkable stream-valley is that immediately to the south of the Traligill, which is called the Allt nan Uamh. It too can boast a disappearing river, which lower down comes gushing out again in full strength from a green hillside, while the cliffs above it possess such varied attractions as a series of caves in which palaeolithic remains have been found, an eagle's nest, and an abundance of *Carex rupestris*. To my mind, however, what is most remarkable about this cliff, and indeed about the whole length of the Allt nan Uamh, is not so much what it does produce as what it does not. *Agropyron donianum* is here conspicuous by its absence. It grows, as I have already said, on the cliff behind the hotel at Inchnadamph, and it extends for a fair distance along the cliff in the direction of the Allt nan Uamh. Where it stops I do not know, because I have never been to look. And so if a botanist who has been visiting the Allt nan Uamh wants an interesting occupation on his walk back to Inchnadamph, I suggest that he follows along the foot of the cliff above the road and observes at what point the *Agropyron* re-enters the picture. It seems strange that it should be so strictly centred on the Traligill valley that it occurs, so far as I have been able to see, in neither of the neighbouring valleys to the south or north.

Seeing how productive is this one relatively small area around Inchnadamph, it might well be supposed that the higher or less accessible mountains of the region would prove even richer; and they certainly do demand further thorough exploration. But it is only fair to add that from my own experience of them they seem to support, for the most part at least, only the dullest vegetation of the dullest hills. Those which I have myself explored, though by no means as thoroughly as they may yet prove to deserve, are Conival, Breabag, Quinag, Stack Polly and Cul Mor; and the whole list of mountain plants from these five hills together would not compare in length with the list from the Inchnadamph limestone. The truth is, of course, that the Torridonian sandstone of which these hills are composed is floristically very poor. But for all that, the recent discovery of *Artemisia norvegica* on that same poor rock (see p. 138) is sufficient indication that these Northern hills, and especially, to my mind, the hills of Coigach, may yet yield hidden treasure.

The journey from Inchnadamph to Bettyhill will inevitably take you (unless, that is, you are prepared to risk the coast road) within easy striking distance of three others of the rarer Scottish plants. Pyramidal bugle (*Ajuga pyramidalis*), which grows on a river bank near Oykell Bridge, is not, as a matter of fact, as uncommon a plant as is often supposed. Not only is it relatively widespread over the northern half of Britain (and it occurs, too, incidentally, on either side of Galway Bay in Ireland), but both in the Hebrides and in Sutherland it is to be seen in numerous scattered localities. I have myself found it in Sutherland, without knowing in advance that it was there, in three quite separate stations, including that near Oykell Bridge, in the course of four consecutive days. It seems to be habitually associated with two particular plants, juniper and bearberry; which means that the best place to look for it is on rather bare, dry, rocky banks and hillsides. Strictly speaking, pyramidal bugle is not a mountain plant in the British Isles, for it is nowhere recorded above 2,000 feet; but it has a mountain distribution in Central Europe, and we have stretched a point to include it. Another rare Scottish plant not strictly within the compass of this book is the sedge *Carex chordorrhiza*, the only known British stations for which are the bogs on either side of Altnaharra. In the great bog at the head of Loch Naver it grows only in the dampest patches, where the most notable of its close associates are *Carex limosa* and many-stemmed spike-rush (*Eleocharis multicaulis*). There seems no good reason why it should not be found elsewhere in Northern Scotland; there are certainly many other bogs where the flora is in general much the same. But it is perhaps worth mentioning that this particular bog must be, to judge at any rate from the presence of such species as dioecious sedge (*Carex dioica*) and few-flowered spike-rush (*Eleocharis pauciflora*) in the flushes around its edges, a good deal less acid than the majority of Scottish peat-bogs. Then finally there is dwarf birch (*Betula nana*), which grows in some quantity both at one corner of Ben Klibreck and on the wet upland plateau which separates Ben Hiel from the main mass of Ben Loyal. The dwarf birch, like pyramidal bugle, is a relatively widespread plant in the northern half of Britain; but it too, again like the bugle, is perhaps most frequent in Sutherland. On Ben Loyal, as also at the head of Glen Clova, it hybridises with the ordinary birch tree (*B. pubescens*), in such a way that, in the former station at least, the nearest colony of the tree is surrounded by quite a dense shrubbery of the hybrid intermediate.

The neighbourhood of Bettyhill, for all its attraction to the botanist, is so palpably not a mountain that it would be wrong, in the present context, to linger there for longer than we need. There are several places on the north coast where the mountain flora descends to the sea-cliffs. Near Durness, for instance, at the western end, purple saxifrage is abundant, while to the east of Thurso there is at least one geo (as the rocky inlets of the sea are there called) which supports a fine colony of alpine saw-wort. But of all such places Bettyhill must be by far the best, and especially one particular rocky hillock, banked up by blown sand, near the mouth of the Naver. At the foot of this hillock, in the damp sand by a little stream, grows the northern sedge, *Carex maritima*; and within a few yards of the sedge, and still at the lowest level on the sand, begins a veritable carpet of mountain avens interspersed with other such calcicoles as globe-flower, yellow mountain saxifrage, purple saxifrage, and dark red helleborine. I should like to be able to append a list of the plants of this remarkable little hill; but unfortunately on my last visit, when I had such a list in mind, the weather was such that on so exposed a spot my courage failed.

Apart from this one spectacular hill there are still three other plants to enhance the attraction of Bettyhill. One, Scottish bird's-eye primrose (*Primula scotica*), hardly falls within the scope of this book. Another, pyramidal bugle, of which there are at least two thriving colonies in the neighbourhood, has already been discussed. But the third, purple oxytropis (*Oxytropis halleri*), which also is here abundant on cliffs, bluffs and banks near the sea, deserves a brief comment to itself. Though primarily, in Scotland, a plant of the coasts—it used to be found, for instance, on the banks of the Firth of Forth opposite Edinburgh—it also occurs, though I know one such station only, in the mountains of the central Scottish Highlands. Here, at an altitude of some 2,000 feet in the neighbourhood of Pitlochry, it is a conspicuous member of a distinguished community, which comprises, among other things, alpine mouse-ear chickweed (*Cerastium alpinum*), alpine milk-vetch (*Astragalus alpinus*) and *Carex rupestris*. *Oxytropis halleri* is one of that small and select class of British plants, of which mossy cyphel (*Cherleria sedoides*) is the most familiar example, which are found in the Alps, the Pyrenees and the Carpathians, but not in the Arctic (see Ch. 3). It has also a fair claim, with its brilliant red-purple flower-spikes rising from among the delicately pinnate silver leaves, to a place in any short list of Britain's loveliest plants. Though it remains in Britain primarily, if anomalously, a plant of the northern coasts, its occurrence near Pitlochry means that in Britain, as on the Continent, it can fairly be numbered also among the plants of the mountains. (For its British distribution see Appendix III, map 5a).

M. C. F. Proctor

Plate XV. A mountain lady's mantle, *Alchemilla glomerulans.* Teesdale

C. D. Pigott

Plate XVI. An alpine lady's mantle, *Alchemilla conjuncta.* Glen Clova

CHAPTER 9

THE MICA-SCHIST MOUNTAINS OF THE CENTRAL HIGHLANDS (J.R.)

(*a*) General Introduction

At this stage in the book, when we come to the flora of the mica-schist mountains of central Scotland, we reach the most difficult problem of all. There can be no doubt that, to do so large and rich an area anything like justice, a whole book should be devoted to it alone. But now that in previous chapters we have accomplished what is virtually a circular tour all around this central area, it may be possible to confine my account of it within reasonable limits. Certain omissions are still unavoidable; but I can only hope that what has gone before may serve to make those omissions less culpable than they would otherwise have been.

The main area which we are now to consider stretches north-eastwards from Ben Lui itself, by way of the ranges on either side of Glen Lochay and Glen Lyon, and then through Ben Vrackie, Ben-y-Gloe and Loch Loch, as far as the two greatest of the great Angus glens, Caenlochan and Glen Doll. And just in case this area is not of itself sufficiently large, there is also an outlying appendage at each end. To the west there is at least one small outcrop of basic rock above Glencoe; to the east there is the short stretch of the Unich Water above Loch Lee.

These mica-schist mountains in general, and especially the Breadalbanes in the centre of the area, carry a vegetation which contrasts sharply with that which we described in detail on the Cairngorms (pp. 112-7). They are extensively sheep-grazed, and in place of the large expanses of heather, crowberry and bilberry which are so characteristic of the Cairngorms the mountain slopes are usually clothed with grassland. On the lower slopes, within the original forest zone, there are often, as there are, for instance, on the familiar south slope of Ben Lawers, rich pastures bright with orchids and mountain pansies. Above the pastures the grassland turf is usually dominated either by fescue (*Festuca ovina* and *F. vivipara*) or—particularly where the soil is damper or snow lies later in early summer—by the mat-grass (*Nardus stricta*). Heather does of course occur on such mountains and may even be locally dominant; but if the whole mountain slope is a heather moor, then it is almost certain that the underlying rock is not a limestone or calcareous mica-schist but some hard acid rock. Anyone who has climbed that noble mountain Schiehallion will have been struck by the contrast in vegetation between it and the neighbouring Lawers range. Indeed the lower heather moor and the upper crowberry-bilberry zone, which we have already described on the Cairngorms, are both here faithfully reproduced. The reason is not far to seek. The main mass of Schiehallion is built of hard acid quartzite, and all the rare plants recorded from the mountain are restricted to the vicinity of narrow limestone belts quite low down on its slopes.

Although the general contrast between basic and acid mountains is thus obvious enough on the slopes, it is less marked in the high mountain vegetation. On exposed ridges in the Lawers range, just as on the Cairngorms, open communities of the sedge *Carex bigelowii* and the moss *Rhacomitrium* occur frequently, while at the other extreme sheltered snow-patch communities are similarly characterised by liverworts, mosses and a few such flowering plants as the least willow (*Salix herbacea*). Even the sparse summit community characterised on the Cairngorms by an abundance of the three-leaved rush (*Juncus trifidus*) has its counterpart on the Breadalbanes. Though on Ben Lawers, for instance, the rush itself is not very common, its place is largely taken, in the scattered vegetation on the summit ridge, by dwarf fescue and spiked woodrush (*Luzula spicata*).

Over the whole of this great expanse of mica-schist there are certain mountain species which are so widespread and abundant that they hardly call for a special expedition at all: they will almost inevitably turn up on a search for one of their rarer associates. And since at least the majority of these species have already been introduced into earlier chapters, I propose at this point merely to list them without comment and thereafter to mention them only incidentally as occasion arises. These, then, are the plants which may be seen on almost any one of the mica-schist mountains or cliffs of Perthshire and Angus :—

alpine meadow-rue (*Thalictrum alpinum*), alpine scurvy-grass (*Cochlearia alpina*), moss campion (*Silene acaulis*), alpine mouse-ear chickweed (*Cerastium alpinum*), sibbaldia (*S. procumbens*), yellow mountain saxifrage (*Saxifraga aizoides*), purple saxifrage (*S. oppositifolia*), mossy saxifrage (*S. hypnoides*), starry saxifrage (*S. stellaris*), roseroot (*Sedum rosea*), chickweed willow-herb (*Epilobium alsinifolium*), alpine willow-herb (*E. anagallidifolium*), dwarf cudweed (*Gnaphalium supinum*), alpine saw-wort (*Saussurea alpina*), mountain azalea (*Loiseleuria procumbens*), alpine bistort (*Polygonum viviparum*), mountain sorrel (*Oxyria digyna*), least willow (*Salix herbacea*), Scottish asphodel (*Tofieldia pusilla*), three-flowered rush (*Juncus triglumis*), three-leaved rush (*J. trifidus*), spiked woodrush (*Luzula spicata*), the sedges, *Carex capillaris, C. vaginata* and *C. bigelowii*, the grass *Poa glauca*, green spleenwort (*Asplenium viride*), holly fern (*Polystichum lonchitis*), and the three clubmosses, *Lycopodium alpinum, L. selago, Selaginella selaginoides*.

I do not wish to suggest that all these plants will be found on the same sort of terrain; some are characteristic of the summits, some of the cliffs, some of the gullies, some of the dry grassy slopes and some of the alpine flushes. But it ought not to take any botanist very many days, whether he is staying at Bridge of Orchy or Tyndrum, at Killin or Fortingal, at Spital of Glenshee or in Glen Clova, before he has seen every plant in this whole long list. Only indeed when he has seen all these does he begin on the species which are exclusively characteristic of the district.

There is also another shorter list of mountain plants which, though hardly falling into the category of the locally ubiquitous, are yet sufficiently widely dispersed over the mica-schist to be characteristic rather of the area as a whole than of any particular part of it. Into this class fall hoary whitlow-grass (*Draba incana*), a plant primarily of fine screes and stony gullies; mountain avens (*Dryas octopetala*) and alpine saxifrage (*S. nivalis*), which prefer respectively the drier and more open, and the damper and more shaded, cliffs and rocks; larger wintergreen (*Pyrola rotundifolia*), a fairly frequent constituent in the vegetation of the most luxuriant cliff-ledges; reticulate willow (*Salix reticulata*), which is often found in the close company of the mountain avens; and finally the sedge *Carex atrata* and alpine meadow-grass (*Poa alpina*), usually plants of the cliffs and gullies, but each liable to stray downwards along the banks of rocky rills. Not only is every one of these seven widely scattered from Ben Lui to Glen Clova; every one of them also, with the possible exception of the willow, is to be found in many other localities outside the area under discussion.

Even when the subject of this chapter has been thus drastically narrowed—and the number of species which henceforth I propose to mention only incidentally is now nearly forty—we are still left with an almost exactly equal number of plants which are especially, if not exclusively, characteristic of the mica-schist. Perhaps the best course at this point, therefore, will be to select certain small but especially interesting areas within the district, and, with particular emphasis on the plants for which each is notable, describe them briefly one after the other. I shall choose, to the exclusion of countless others hardly less rich, just ten such small areas, which I shall treat in an order from west to east; and if some of the descriptions show an irritating lack of local precision, that arises from the desire to afford the plants which grow in these places every possible protection from their enemies the collectors. When finally these descriptions are concluded, any plants which have not received their due mention can be separately discussed at the end of the chapter.

(*b*) Argyll and Perth

The first of my selected areas is a small outcrop of basic rock, to which I have already referred, in the mountains above Glencoe. Whatever the geologists may say about the composition of this rock—and it is marked on the Geological Survey map as limestone—it has to the layman precisely the appearance of mica-schist. Half way along a towering north-facing buttress there is a sudden and obvious change in the rock. On one side, as can be seen from a distance of two or three miles, there is the usual dark slabby rock of Glencoe, so beloved of climbers but for botanists so poor in all but hawkweeds; on the other side are the broken, shelving, overgrown ledges that are typical rather of Ben Lui or Ben Lawers than of Glencoe. In the middle of this rich stretch a deep chasm scars the face of the cliff, with a small stream splashing down it in a chain of falls. On the way up to this chasm several of the plants which I have for present purposes classed as locally ubiquitous already come into sight—chickweed willow-herb, for instance, by the rills, *Sibbaldia* in the drier grass, and on the fine screes over to the right the alpine mouse-ear chickweed. But my immediate concern is only with the short stretch of the stream, a mere

M. C. F. Proctor

Plate XVII. Sibbaldia procumbens. Ben Nevis

fifty yards or so in length, where it first emerges from the almost perpendicular cliff to flow, at the bottom of a small ravine, through two deep pools with a low fall between. At the head of each pool, on the true left bank, is a sheer face of rock, pitted here and there with little pockets where presumably the rock has weathered away, while lower down the pool the rock gives way to a steep stony bank. This is the home of two rare and interesting plants. In the pockets of the sheer rock grows drooping saxifrage (*Saxifraga cernua*); on the steep stony banks, accompanied by the oak fern (*Thelypteris dryopteris*) which does its best to camouflage it, is mountain bladder-fern (*Cystopteris montana*).

This remarkable locality has only very recently become known to botanists. It was discovered in 1949 (as the result, I believe, of a study of the geological map) by Messrs. E. C. Wallace and P. R. Bell. Until then, the statement about *Saxifraga cernua* in Bentham and Hooker was generally thought to contain the whole truth; "In Britain, only known on the summit of Ben Lawers, where, however, it very seldom flowers, and is now almost extinct". Whether or not the impression conveyed by the last two words is strictly accurate (see p. 85), it is at any rate consoling to know that so rare a plant has a second British station. In this second station, too, it seems to flower more freely than on Ben Lawers. By the time of the year when Max Walters and I visited it, towards the end of August, 1953, the flowers had withered; but it was evident that a fair proportion of the plants, besides producing the usual little red bulbils in the axils of the upper leaves, had condescended to flower as well.

The mountain bladder-fern is not so scarce a plant as the saxifrage, but it is by no means common either. Its favourite habitat is a damp mossy ledge, invariably, in my experience, with a northerly exposure (Pl. XXII, p. 181). Locally abundant on Ben Lui, and in at least one corrie to the south of Glen Lochay, it is also scattered here and there, usually in small colonies, not only over the hills between Glens Lochay and Lyon, but also, apparently, as far afield as Helvellyn in one direction and the hills of Skye in another. Personally, however, though I have seen it in three separate spots on Ben Lawers and in

---

Plate XVIII. Two mountain grasses
a. (*above*) *Poa glauca,* Ben Douran, Argyll. (*M. C. F. Proctor*)
b. (*below*) *Agropyron donianum,* Inchnadamph, Sutherland. (*J. E. Raven*)

several others to the west, I have never yet seen it further to the east. There are plenty of suitable places for it in Angus, especially on the Caenlochan cliffs (where it has in fact been recorded) and in Glen Fee. But it seems to be one of those plants with so definite a preference for the Western Highlands—it even occurs, for instance, in the same presumably acid corrie near Ben Nevis as *Carex lachenalii*—that if it is still to be found at all to the east of Perthshire it is likely to take a lot of finding. Incidentally, it seems in most of its localities every bit as shy about 'fruiting' as is *Saxifraga cernua* about flowering. Until I saw it in its Glencoe station I had never succeeded in finding a single fertile frond. But in this one place at least, where it is unusually large and vigorous, it overcomes its shyness and produces spores (or at any rate did in 1953) quite freely.

This first of my chosen areas is still so little known to botanists that the following field-note may be of interest. It was written on the spot on our visit in August, 1953: "Altitude c. 2,700 ft. *Saxifraga cernua*: 3 plants with bulbils, several smaller, scattered over c. 2 sq. yds. of ± N-facing damp vertical rock-face by stream; associated spp. of higher plants very few: *Oxyria, Cystopteris fragilis, Saxifraga aizoides, Sedum rosea.* Several bryophyte patches, but much bare rock."

My second area, like my first, is a fierce north-facing mountain cliff, this time on one of the hills near the head of Glen Lyon; and though this second area is floristically even richer than the first, I am primarily concerned in this case also with only two particular species, namely chestnut rush (*Juncus castaneus*) and the sedge *Carex atrofusca.* Apart from these two, the cliff is notable especially for *Bartsia alpina*, which here, as on several other hills in the neighbourhood, grows freely on the damper parts of the rock. But since *Bartsia* is an even more prominent feature of the next of my selected areas, I shall reserve such comments as I have to make about it until later.

Chestnut rush is another of those plants, like mountain bladder-fern, with an apparent preference for the western half of the mica-schist mountains. True, it used to be found near the head of Glen Doll and is still to be seen, in small quantity, on the boggy slopes above the Caenlochan cliffs. But again, if you want to see it in comparative abundance, then you should look for it on the mica-schist hills from Ben Lawers itself westwards. From Ben Lawers to Ben Lui any mica-ceous bog from an altitude of about 2,500 feet upwards may prove to contain it; but the surest way to find it is to follow a stream up from that altitude to its source. That at any rate is how I have myself usually discovered it, not only on such hills within the area as Meall nan Tarmachan or Ben Douran, but even on so relatively barren a mountain as Sgurr na Ciche at the head of Loch Nevis. Its stations near the cliff under discussion confirm this advice. There are a number of little trickles down the face of the cliff, the edges of which are the favourite habitat of *Bartsia alpina*. But when these trickles leave the cliff proper for the very steep micaceous slope at its foot, the bartsia ceases abruptly and its place is taken by chestnut rush. I doubt if there is any locality in Britain where the rush is half so abundant; and its flowering season seems, here at least, to be so protracted—probably because the snow lies longer at some points than at others—that in late July it is to be found in all stages from fresh flower to ripe fruit.

If the rush in this locality is the counterpart of the bladder-fern in the other, the sedge *Carex atrofusca* corresponds to the drooping saxifrage. For it also has only one other known British station, which too is on (or at any rate very near) Ben Lawers. In the Lawers station (where it was first discovered by George Don, but, like so many of his best finds, subsequently discredited), it grows on damp stony patches on the watershed between two valleys and on the boggy slopes beside the rills which flow from these into Glen Lyon to the north. Its associates in that locality will be discussed later. On the cliff further to the west its habitat is somewhat different. It here colonises, in striking profusion, the broader grassy ledges up the face of the cliff. No doubt all these ledges are, like its other station, perpetually pretty damp; there is, as I have said, plenty of water trickling down the face of the cliff. But it is curious, when the two localities are compared, how little inclination the sedge here shows to join *Juncus castaneus* on the wetter parts of the steep grassy slope below the cliff.

The third selected area, a few miles east-south-east from the second, is at the head of one of the several burns that feed the Lochay from the south. Here, where it first spills itself from the upland plateau, the young stream flows for a while, in summer often as the tiniest trickle, down a stony gully at the foot of a shelving rock; and for the botanist not only the rock above it but the stony gully as well is full of interest. The gully itself is the best place I know to study in the field that small but often misunderstood group of plants, the alpine pearlworts. The cliff can yield, among many of the more widespread plants such as mountain avens or reticulate willow, not only the bartsia to which I have already referred, but also two of the rarest of Perthshire plants, rock whitlow-grass (*Draba rupestris*) and woolly willow (*Salix lanata*). All these indeed grow within a few yards of one another; and if we care to extend our range for half a mile or so, to the top of the spur that forms the eastern flank of the valley, then we can add yet one more to the list of local rarities in the shape of alpine rush (*Juncus alpinus*). Few places in the Highlands can better repay so relatively short a walk; and anyone who happens, because of my deliberate vagueness, to go up a parallel valley by mistake will very probably be rewarded instead by the sight of an unusual abundance of mountain bladder-fern.

The difficulty about the alpine pearlworts is that, though *Sagina saginoides* is, typically, distinct enough from the abundant *S. procumbens,* there is also a third plant, now classified as a subspecies of *S. saginoides* but formerly given specific rank under the name *S. scotica,* which is almost exactly intermediate between the two; and needless to say, the three have a habit of growing, as they do in the gully in question, mixed up together. On and near Ben Lawers the situation is still further complicated by the presence of yet another allied species, *S. intermedia,* which too, as can be seen in the gullies on the north side of Beinn Ghlas, is primarily a plant of fine screes; and what is more, *S. intermedia* also seems to me to have two forms, one with glandular pedicels and mucronate leaves, the other with glabrous pedicels and blunt leaves. And finally, to make matters yet worse, there is even a mountain form of the sea pearlwort (*S. maritima*), which is said to occur, notably on Ben Nevis, at altitudes up to 4,000 feet. But to the south of Glen Lochay these last two complications are, to the best of my knowledge, absent: the gully under discussion is content with a mixture of *S. saginoides, S. procumbens* and the intermediate. I was fortunate on my last visit to be in the company of Professor Clapham, who was at that time working on the genus for the new Flora. From my own limited knowledge of the behaviour of the intermediate plant I should feel inclined to accept his final tentative observation that it "may be the hybrid *S. saginoides* x *procumbens,* as suggested by its morphological features and many abortive capsules."

The interest of *Bartsia alpina,* which here, as in the last area discussed, grows in fair plenty on the damper parts of the cliff, lies chiefly in its British distribution (see Appendix III, Map 2b). There are two areas in Britain in which it is not very uncommon, namely the limestone districts of North-Western England and the mica-schist mountains of Perth and Argyll; elsewhere in the British Isles it is unknown. There are in fact two peculiar features involved here. In the first place it provides yet another example of a plant which evidently has a marked preference for the west. Once again there are numerous localities in Angus where it would seem, to judge from its behaviour in the Breadalbanes, to have everything it could possibly require; yet it has never been recorded from Angus, and almost certainly now never will be. And in the second place, whereas in the north of England it is often, as it is in Upper Teesdale for instance, a plant of damp meadows on the fringes of the hills, in Scotland it seems nearly always to prefer the mountain cliffs. There is admittedly one notable exception to this latter generalisation: the meadows at the foot of Ben Lawers are a well known locality for bartsia. But the peculiar fact remains that a plant which in England seems capable of contending with pretty vigorous competition, in Scotland tends to take rather to places where the competition is greatly reduced. There must presumably be some explanation of this fact, as also of the apparent preference of the plant for the west. But, so far as I know, no attempt at an explanation has yet been forthcoming, and it is certainly not for me to repair the omission.

Woolly willow (*Salix lanata*) (Pl. XX, p. 177), which in this one locality grows with the bartsia, none the less provides in general a striking distributional contrast. Its headquarters in Britain is unquestionably at the eastern end of the mica-schist. In Glen Fee for instance, though it is very local and in one of its stations so intermixed with another mountain willow, *S. lapponum*, that there is little of the pure strain left, it is in tolerable plenty; and it recurs, in smaller quantity, both near the head of Glen Callater, over the county boundary into South Aberdeenshire, and on one rocky ridge, possibly the best place in all Scotland for variety of mountain willows in pure form, above Caenlochan. In Perthshire, on the other hand, it is excessively rare; though I have seen near Ben Lawers a hybrid of which *S. lanata* was probably one parent, I have found the pure strain only on the cliff in question, where it is by no means abundant. In Argyll, or for the matter of that any other western county, it is unknown. It seems likely therefore that this one small cliff to the south of Glen Lochay is the only place in Britain where these two mountain species, with their complementary distributions, can be seen growing together.

Rock whitlow-grass, the last of the specialities of the cliff, is another exceedingly rare and local plant, but its British distribution presents a pattern of a different type. It grows in this particular station on broad cliff-ledges near the top of the gully; and though there are not many plants of it, probably not more than twenty in all, they are, compared with the majority of those on Ben Lawers itself, very fine and large. Outside Perthshire (and apart from one mountain in Argyll which really belongs to the Perthshire group) it grows also, as I mentioned in an earlier chapter, on one or two patches of basic rock in the Cairngorms, and on one rich but in these days neglected mountain of the far north, Ben Hope. On Ben Hope, incidentally, it is said to grow with another plant characteristic of the mica-schist, alpine sandwort (*Minuartia rubella*). Both, according to the *Comital Flora,* attain on Ben Hope, as on Ben Lawers, an altitude of 3,980 feet. According to the Ordnance Survey map, on the other hand, Ben Hope is only 3,040 feet high. But there is this much justification for the version in the *Comital Flora,* that rock whitlow-grass is generally a plant not only of the richer rocks but of the richer rocks on the higher hills. In its Glen Lochay station, at an altitude of little over 2,500 feet, it is probably as low as anywhere in Britain.

The next area that I want to describe is a stretch of east-facing cliff, some 150 yards in length and at an altitude of about 2,250 feet, on a spur of Meall nan Tarmachan. This cliff again is the station for a number of rare and interesting plants, including a sedge, *Carex rupestris,* a grass, *Agropyron donianum* and a fern, *Woodsia alpina.* But it is to my mind even more notable for the extraordinary luxuriance of its vegetation. A full list of the plants to be found here would occupy a whole page or more, and the list that I shall shortly give is therefore only of the vegetation of a few square yards. But there are several species not figuring in that list which deserve a mention. First there is the willow, *Salix arbuscula,* which here, as on several other cliffs in the near neighbourhood, ousts *S. reticulata* from its normal position in the Breadalbanes of commonest of the willows of the mountain cliffs. Then there are the two wintergreens, *Pyrola rotundifolia* and *Ramischia secunda,* which are scattered here and there in small colonies on the damper ledges. Then, on the drier and more exposed ledges, there are several of the more showy of mountain hawkweeds, especially one which is to my mind the most beautiful of them all, *H. sinuans*. And finally, and perhaps most striking, there is a whole group of plants which, to a botanist who knew England but not Scotland, might seem characteristic either of the hay-meadows of a district such as Upper Teesdale or of a woodland drive. Globe-flower (*Trollius europaeus*), wood cranesbill (*Geranium sylvaticum*) and water avens (*Geum rivale*) represent the first group; primroses, wood-anemones, red campion (*Melandrium rubrum*), dog's mercury (*Mercurialis perennis*), and even the wood vetch (*Vicia sylvatica*) represent the latter. I know of only one cliff in Britain, a few miles to the north-east of this one, which can vie with it in the luxuriance and variety of its vegetation; and that other is also a locality for both *Agropyron donianum* and *Woodsia alpina*. Elsewhere in Europe, however, a very similar vegetation is to be seen, I believe, in the Scandinavian birch woods. If only it were possible to surround this cliff and the area below it with a fence which would exclude both deer and sheep, it seems most likely that this same vegetation, including the dwarf willows, would soon invade the slopes beneath the cliff. Certainly it survives where it now is, on the shelving ledges, only because of its relative inaccessibility to animals.

Both *Carex rupestris* and *Agropyron donianum* have already been mentioned several times in an earlier chapter (Ch. 7.); for at Inchnadamph, as here, they frequent the same kind of cliffs. The sedge, indeed, is very much more abundant on the north-western limestones than it is in central Scotland. In each of the three stations where I have myself seen it on the mica-schist, it grows on the barer parts of a cliff, in small quantity only, and it does not usually seem to flower at all freely. The rare grass, on the other hand, is on this one cliff locally abundant, and the reason for which it is not very often seen is probably only because, in a normal season, it does not come into full and conspicuous flower until August or even September. The following description of a typical colony, which I quote verbatim for its list of associated species, is based on a visit as late in the year as September 4th, when the grass was still in such beautifully fresh condition that we wondered whether it ever succeeds, in its mountain stations, in ripening its seed:—

"Wet, more or less vertical gully, facing S.S.E., between two bare, vertical rock faces; water seeping through and over a rich vegetation. *Agropyron donianum* was here frequent, with the following associated species:—

Agrostis tenuis	(bent)
Alchemilla alpina	alpine lady's mantle
Alchemilla glabra	(lady's mantle)
Angelica sylvestris	angelica
Cerastium alpinum	alpine mouse-ear chickweed
Dactylis glomerata	cock's foot
Euphrasia sp.	(eyebright)
Filipendula ulmaria	meadow-sweet
Festuca rubra *Festuca tenuifolia* }	(fescue)
Hieracium prenanthoides	(hawkweed)
Holcus lanatus	Yorkshire fog
Linum catharticum	purging flax
Luzula sylvatica (few large clumps)	greater wood-rush
Plantago lanceolata	ribwort plantain
Poa nemoralis	(meadow-grass)
Ranunculus acris	meadow buttercup
Rhinanthus stenophyllus	(yellow rattle)
Rumex acetosa	sheep's sorrel
Salix nigricans (one large bush)	(willow)
Saxifraga aizoides	yellow mountain saxifrage
Viola riviniana	dog violet

On the bare, somewhat drier rock were also :—

Draba incana	hoary whitlow-grass
Pinguicula vulgaris	butterwort
Poa pratensis	meadow-grass
Silene acaulis	moss campion
Thymus drucei	wild thyme"

I would like to cite also, but space forbids, the list of associated species in what is probably Don's original locality for the grass, the other luxuriant cliff to the north-east. It contains, in addition to several plants common to the two sites, also such diverse species as ladies' fingers (*Anthyllis vulneraria*) and bird's foot trefoil (*Lotus corniculatus*), purple saxifrage (*S. oppositifolia*) and alpine forget-me-not (*Myosotis alpestris*).

The two British species of *Woodsia*, *W. ilvensis* and *W. alpina*, are apparently distributed in Britain over almost exactly the same range. It may well be that in Snowdonia, where they certainly both occur, there is about as much (or as little) of the one as of the other; but in Scotland, or at any rate on the mica-schist, I am convinced from my own experience that *W. alpina* is much the less rare. It and the mountain bladder-fern tend to be found on the same mountains, if not on the same cliffs; both are to be found, or have at one time or another been found, on Ben Lui to the west, Ben Lawers in the centre and Caenlochan to the east. But there the similarity in their preferences ends. Whereas, as I said earlier, the bladder-fern likes a damp, mossy, often overhung ledge with a northern exposure, the *Woodsia* prefers a very narrow crevice in an open rock and does not seem to mind which direction it faces. The same is true also of *W. ilvensis*; in the one and only locality on the mica-schist where I have yet succeeded in finding it, in Glen Fee, about a dozen plants of it are so tightly wedged side by side in a narrow crevice of a south-east-facing rock that they leave no room at all for potential competitors. Both species of *Woodsia*, of course, in common with all the other rarer British ferns, have been the victims of the unscrupulous collecting of the nineteenth century. It may be that in a number of its old Scottish stations *W. ilvensis* has been totally eradicated. But I can but wonder—and the Glen Fee plant tends to strengthen my doubt—whether in Scotland the line of demarcation between the two species is not a good deal less clear than it is in Wales, with the result that a number of the older records are mistaken.

Next in our progress from west to east we come inevitably to Ben Lawers itself; and even if two of my chosen areas are in its close neighbourhood, yet the mountain is so well known to botanists, and so much has been written about its flora, that I shall pause there no longer than I need. On and around the summit of Ben Lawers are congregated so many of the rarest mountain plants that there is little to be done in a necessarily short account beyond merely listing them. If it were possible to be conveyed to the summit cairn and to start botanising from there downwards, the first interesting plant to be encountered would be either the pearlwort *Sagina intermedia* or the rock whitlow-grass, both of which grow, if not on the cairn itself, at least on the broken rocky ridge around it. Next we should come upon drooping saxifrage and alpine forget-me-not, the former doing its best to hide itself in little hollows between the tumbled rocks, the latter flaunting its sky-blue flowers (and it is a Mediterranean sky at that) on the vertical sides of the largest of these boulders. Then, if we chose to walk along one of the ridges, we ought soon to find the two-flowered rush (*Juncus biglumis*), growing in a damp micaceous hollow just below the crest of the ridge; and with it would be not only its commoner relative, *J. triglumis*, but also, very probably, both Scottish scurvy-grass (*Cochlearia micacea*) and the puzzling form of mouse-ear chickweed (*Cerastium*), so widespread on Ben Lawers, which is thought to be the hybrid *C. alpinum* x *vulgatum*. If, on the other hand, we chose rather to plunge down into one of the corries below, then we should probably find instead either the sedge *Carex saxatilis* and the chestnut rush, which thrive around the springs on grassy slopes, or, if we preferred to go down one of the stony gullies between the cliffs, not only the plants which we have already met on the summit, but also, on the cliff-ledges beside the gully, such a variety of beauties as alpine cinquefoil, boreal fleabane (*Erigeron borealis*, Pl. 14, p. 179), rock speedwell (*Veronica fruticans*, Pl. 16, p. 191) and perhaps also alpine speedwell (*V. alpina*). And when, finally, we emerged from the gully on to the "apron" of sloping rock and fine screes, we ought soon enough, if we have chosen the right cliff at least, to come upon alpine sandwort (*Minuartia rubella*) and snow gentian (*Gentiana nivalis*). It is no wonder that, for a century and a half, Ben Lawers has been the Mecca, not only of field botanists, but also, unfortunately, of alpine gardeners; and it is a tribute to the wealth of its flora that, despite constant depredations, none of its rarities has yet been exterminated. Some of them doubtless take a lot more finding now than of old: for my own part I had to climb the mountain at least three times before I had seen all the plants named in this paragraph. And there is still, after more visits than I care to compute, one of its treasures which I have never yet seen on Ben Lawers, brook saxifrage (*S. rivularis*). But even in that case I have every reason to believe, as in every other case I know from my own experience, that somewhere on the mountain a colony of the plant is still surviving.

Only three of the plants I have just mentioned seem to call at the moment for more detailed comment. The forget-me-not, which is still, for all its desirability in the rock-garden, remarkably abundant on Ben Lawers and its neighbours, is elsewhere in Britain native in only one small area of Upper Teesdale in the vicinity of Mickle Fell (see p. 143). It has also been recorded in the past from the Caenlochan cliffs; but there is good reason to suppose, as we shall see in due time, that it was deliberately sown in that station, and it has not, so far as I know, been seen there for several years. Alpine speedwell, on the other hand, is not by any means a common plant on Ben Lawers, where I can only remember seeing it on one stretch of cliff some distance away from the summit. There is, moreover, a trap here for the unwary. The plant which used to be called *Veronica humifusa*, but has now been demoted to *V. serpyllifolia* subspecies *humifusa*, is abundant on Ben Lawers, as also in parts of Glen Clova, in the damp patches around mountain flushes and rills; and it often contrives to look, to a superficial glance, misleadingly like *V. alpina*. As for *Gentiana nivalis*, the most notable fact about it, apart perhaps from its colour, is that it is one of the very few British mountain plants (the only others being the mountain species of *Euphrasia* and *Rhinanthus* and *Koenigia islandica*) which are annuals, not perennials. In consequence it varies very greatly in size, both from place to place and from year to year; some specimens are a mere inch or so high and bear only one or two flowers, others, especially on one particular cliff ledge on Ben Lawers where they are accompanied by the fleabane, attain a height of four or five inches and as many as sixteen to twenty flowers. The flowers themselves only open, unfortunately, in direct sunlight, when they are just as dazzlingly blue, though not nearly so large, as those of spring gentian (*G. verna*). In dull weather the smaller plants are not at all easy to see.

Apart from the summit and the cliffs below it, the richest part of the Ben Lawers group is around the source of one of the streams that flow north into the Lyon. This is, as a matter of fact, the small area to which I have already referred in this chapter in connection with the very rare sedge *Carex atrofusca*; but *Carex atrofusca* is by no means the only rare plant for which it is noted. The actual source of the burn consists of a large wet patch of peat and stones, at the foot of a very steep slope and at an altitude of about 2,700 feet. As soon as it leaves this patch, the burn drops pretty rapidly, between damp grassy slopes, into the valley to the north. The area with which I am concerned is only the flat patch itself and the grassy slopes for a hundred feet or so below; for though this is an area which might seem at first sight to be nothing out of the ordinary, it is floristically unique in Britain. The wet stony patch carries a sparse enough vegetation, but among its constitutents are the two-flowered rush and the sedges *Carex microglochin* and *C. atrofusca*; and on the grassy slopes just below, these three are joined by the chestnut rush and *Kobresia simpliciuscula*.

Carex microglochin was added to the list of British mountain species by Lady Davy and Miss Gertrude Bacon (later Mrs. Foggitt). They found it, in this locality, on July 25th, 1923, and they sent a specimen, for confirmation of their tentative identification, to Dr. G. C. Druce. With his usual indefatigable enthusiasm Dr. Druce made a special journey to Scotland to see it, and his visit to the locality, made just a month later, resulted in a description (in the Botanical Exchange Club's Report for that year) from which I shall quote a few sentences:—

"It grows on the grassy mountain slopes in places where the runnels open out in flattish spaces, forming micaceous shallow bogs where the vegetation is sparse and small. It avoids the more sloping and grassy places over which the water trickles, and in which *Carex capillaris* and *C. saxatilis* delight to grow. The latter sedge occurs in a small and scattered form, but it also does grow with *microglochin,* which has as its associates *Carex dioica, C. atrofusca,* a small form of *C. flava, C. nigra, Juncus triglumis,* and rarely *Kobresia.* In such habitats *Carex microglochin* was tracked up to an altitude of 3,000 feet. A few specimens were found as tall as five inches, but usually they are from three to four inches only. On the slopes between these micaceous bog-terraces grow *Silene acaulis, Saxifraga oppositifolia, Juncus biglumis, Luzula spicata, Sagina saginoides,* and other alpines now mostly over flower. The higher rocks show *Salix herbacea* and *S. reticulata, Cerastium alpinum, Athyrium alpestre, Polystichum lonchitis;* but the *Hieracia* were over flower or non-existent."

Not only is this an excellent description of this little area, but it also raises a few points which need making before we move on eastward. In connection, first, with *Carex microglochin* itself, a recent pH test by Dr. Elizabeth Davies of a soil sample from a bog in which the sedge was growing showed that the soil had a pH of 7.05. "This", as Dr. Davies remarks (Davies, E., 1953), "is an interesting contrast to the acid bog areas where one normally finds *Carex pauciflora* (the species that resembles *C. microglochin* most closely in Britain), which in my limited experience always grows on soils with a pH varying between 4·5 and 5·5". Then, in connection with *Kobresia simpliciuscula,* a plant I have hardly had occasion to mention before, it is perhaps worth remarking how very similar is its distribution in Britain to that of *Bartsia alpina,* which I have already discussed. *Kobresia* is confined to almost precisely the same two widely separated areas as the bartsia. Both are to be found, and in a fair number of stations, in Yorkshire, Durham and Westmorland, Perth and Argyll; neither are to be found in any other county, not even Angus. And finally, in connection with the plant regarded by Druce as dioecious sedge (*Carex dioica*), it is certainly worth drawing attention to another recent paper in *Watsonia*, this time by Mr. N. Y. Sandwith (1952), in which a fascinating problem is posed. It appears that besides typical *C. dioica,* which is a rare plant on Ben Lawers above about 2,000 feet, there is also another plant, never before described, somewhat intermediate between *C. dioica* and the Scandinavian species, unknown in Britain, *C. parallela,* but with "more characters in common with *C. parallela* than with *C. dioica*". Whether the plant which accompanies *C. microglochin* is the typical *C. dioica* or this newly described intermediate, I do not myself know; and in any case, as Mr. Sandwith says in conclusion, "the status of this puzzling sedge is by no means clear; but I think that a preliminary note was required to call attention to it as an interesting subject for serious investigation."

(*c*) Angus

The next step in our eastward progress is a long one. I have never been, I am sorry to say, either on Ben-y-Gloe itself or on the steep and rocky slopes above Loch Loch. The whole of that area deserves a thorough exploration. The yellow-flowered *Oxytropis campestris,* which is known elsewhere in Britain only in Glen Fee (see p. 179), is certainly still to be seen near Loch Loch, while Ben-y-Gloe may yet yield an even greater treasure in *Rubus arcticus.* This relative of the cloudberry seems lately to have been rather too definitely excluded from the British list. According to Smith's *English Botany* "the late Rev. Dr. Walker, Professor of Natural History at Edinburgh, informs me in the year 1782 of his having gathered this beautiful plant in rocky mountainous parts of the Isle of Mull. Mr. Sowerby has been favoured by Richard Cotton, Esq., with dry wild specimens from the high regions of Ben y Glo, Blair, in Scotland, which agree with that in our plate." Again, the Report of the Botanical Society of Edinburgh for 1841 records "a specimen, too imperfect to decide, by J. Robertson, from head of Glen Tilt", no great distance from Ben-y-Gloe. Druce (1920), in his article on "The Dubious Plants of Britain," asks: "May the dwarf alpine one-flowered form of *Geum rivale* have been mistaken for it?" But not only does it seem unlikely that so familiar a British species should have been mistaken, and by competent botanists too, for one so unfamiliar, but there are also, in the herbarium of the British Museum, specimens which were apparently collected in Scotland and which are undoubtedly rightly named. It is hard to resist the conclusion that *Rubus arcticus* will one day be reinstated in our list, and the exploration of Ben-y-Gloe stands high on my agenda. But since all that I know of the mountain at present is at best second-hand we must move on eastwards to my next selected area, in the neighbourhood of Caenlochan Glen.

The flora of the Caenlochan cliffs is, I suppose, second only to that of Ben Lawers itself, and it includes two or three mountain species which have never been recorded from the Lawers group of hills. But before we turn to the cliffs, we should consider first four plants which are to my mind especially characteristic of the great rolling upland plateau around the head of the glen. For even if all of the four are occasionally encountered in the Breadalbanes (and two of them at least are actually tolerably widespread there), there can still be little doubt that it is in the zone which we are just entering that they are most at home. This upland plateau is a strange and inhospitable place. Stretching for mile after mile at an altitude round about 3,000 feet, it consists quite largely of bare and forbidding peat-hags interspersed with tussocks of heather. But here and there in the district, barren as much of it is, are little areas of exceptional interest. One, the area of serpentine outcrops on which red alpine catchfly (*Viscaria alpina*) grows, has been discussed in an earlier chapter (pp. 108-9). I am concerned now with two others of a very different type, one a semicircular amphitheatre of scree topped by a low wall of rock, the other a little hollow in the peat which is always so damp as often to contain a shallow pool.

The first of these two localities, the amphitheatre of scree, is the best place I know for comparing the two most attractive of British mountain grasses, *Alopecurus alpinus* and *Phleum commutatum* (Pl. VIIa, p. 112). Half way down the steep slope is a ring of green alpine flushes, each of which gives rise to a small runnel; and most of these runnels spread out in their downward course to form little streams, perhaps a yard wide but barely an inch in depth. These streams provide the favourite habitat of both the grasses; but though the two species often grow within a few feet of one another, they almost always reveal a slight difference of taste. The *Phleum* (the awns of which, when it is fresh, give it a beautiful red-purple sheen) grows among other grasses and sedges on the streams' banks; the *Alopecurus* (the awns of which are much shorter and which is rather the colour of a thunder-cloud) prefers to grow away from much competition in the middle of the shallow stream. In this particular station the two are associated (as also, I believe, in the Cairngorms) with alpine speedwell and the very dwarf mountain form of the kingcup (*Caltha palustris* ssp. *minor*)—not to mention alpine willow-herb (*Epilobium anagallidifolium*) and other widespread plants of mountain flushes.

The shallow depression in the peat is, as one might expect, the habitat of mountain sedges rather than grasses, and the two most notable of its inhabitants are *Carex aquatilis* and *Carex rariflora*. Both are interesting plants. *Carex aquatilis* is primarily a plant of the lower ground in hilly districts, and it occurs, chiefly on the marshy banks of rivers and lakes, as far to the south as Pembrokeshire. In Scotland it has two very different kinds of station, between which there is usually a considerable gap where it is not to be seen, and its stature and general appearance is widely different in the two. In the valleys, as for instance in the lower reaches of Glen Clova itself, it is tall, upstanding and leafy. On the upland plateaux, usually at altitudes exceeding 2,500 feet and in situations similar to this boggy hollow, it takes on a very dwarf and condensed form, and is often not only bent over to one side but also spirally twisted. There has in the past been a great deal of discussion of this upland form; it has been given a number of varietal names and even credited with hybrid origin. But it is now generally held that, for all its different appearance, it is conspecific with the plant of lowland river banks.

Carex rariflora, on the other hand, is not only primarily but, in Britain at least, exclusively a plant of the higher uplands; and though it has one or two isolated stations as far to the west as the western fringes of Perthshire, it is still pre-eminently a plant of the Eastern Highlands (see Appendix III, Map 3b). It has one curious characteristic which I have not noticed in any other British sedge: though I have never read or heard any comments to the same effect, I know for certain from my own experience that in some years it hardly flowers at all. In this very area, indeed, on our first expedition in search of it, my father and I spent several fruitless hours examining just the sort of marshy hollow that it likes, and not a single sign of it did we see. Yet on our second visit to the same area, a year or two later, at least one of the very hollows which we had so thoroughly searched bore scores of fruiting spikes. I have observed the same habit on Lochnager too, where in some years it is apparently absent, in others locally common. I suspect that this is one reason why in Perthshire, where it is anyhow very much less widespread, it seems only to be noticed at fairly infrequent intervals.

As for the Caenlochan cliffs, they also, like Ben Lawers, have been so often and so thoroughly explored and described that no detailed account of them is needed here. At the south-western angle of the cliff, where there is a long red shoot of porphyry, there is an abrupt change in the rock; on the one side is granite, on the other mica-schist. Needless to say, the mica-schist is far the richer; blue sow-thistle (*Cicerbita alpina*) is the only rarity which prefers the granite to the mica-schist, while the rarest plant which seems almost equally at home on either is *Poa glauca*. Of the rarer plants of the Caenlochan mica-schist we have met with the majority earlier in this chapter—boreal fleabane, for instance, snow gentian, rock speedwell, woolly willow and *Woodsia alpina*; and I shall therefore say no more about these than that they are to be treated here with at least as much care and respect as in their other localities which I have described. But there are also two other plants native in Caenlochan which call for slightly longer comment, namely alpine penny-cress (*Thlaspi alpestre*) and alpine meadow-grass (*Poa alpina*).

Alpine penny-cress presents a pattern of distribution in Britain different from any plant I have yet discussed. Widely dispersed on the limestone hills of England and Wales, from the Mendips in the south to Teesdale in the north, and occurring also in both Snowdonia and the Lake District, it is in Scotland one of the rarest of mountain plants. Personally I have seen it in three Scottish localities only, and not many more than three are known. In Caenlochan, it chooses mainly the upper parts of the very steep stony gullies, ascending to 2,900 feet. On Beinn a Chuallaich, in Perthshire, it occupies a small outcrop of basic rock, on an otherwise dull mountain, at an altitude of about 2,000 feet, and is accompanied only by such relatively common mountain species as purple saxifrage and Scottish asphodel. On Fionchra in Rhum, which I have already mentioned in an earlier chapter, it grows on outcrops of mugearite at only 1,250 feet, while in its other locality on the same island, where I have not seen it, it

R. M. Adam

Plate XIX. Habitat of a mountain willow, *Salix lapponum.* Lawers range

M. C. F. Proctor

Plate XX. Woolly willow, *Salix lanata*. Clova Mountains

drops below the 1,000-foot contour. Since on the Mendips and in North Wales it is hardly to be ranked as a mountain plant at all, it is very evident that it is more particular about the rock on which it grows than about its altitude. As for the various species, subspecies or forms into which the British plant has been split, it seems to me that, until they are better understood, it is best for practical purposes to regard them as mere varieties of a single polymorphic species. The Caenlochan plant, at all events, fits satisfactorily into none of the forms described and named.

The only interesting thing about *Poa alpina* in Caenlochan is that these cliffs are one of the two localities known in Britain where the plant is not always viviparous (cf. p. 141).

I cannot leave Caenlochan Glen without a short quotation, the relevance of which will be immediately obvious. In the *Journal of Botany* for 1885 the following communication appeared over the signature of F. Buchanan White:—"I notice in the Report for 1883 of the Botanical Record Club that *Myosotis alpestris* has been gathered in Caenlochan, Forfarshire, by the Revs. H. E. Fox and E. F. Linton. I fear, however, that the plant is not indigenous in this locality, as, unless my memory deceives me, I was told by a horticultural friend that he had sown a quantity of the seed—derived from cultivated Ben Lawers plants—along with the seeds of other alpines, in Caenlochan. One result of this was the discovery there, by some members of the Edinburgh Botanical Society, in 1880, of *Myosotis alpestris, Erinus alpinus, Primula auricula,* and other aliens. I quite believe that my friend was in no degree aware of the reprehensibility of his proceedings, or that if his name was revealed it would be held in execration by all British botanists. I am sorry to say that he also sowed (unless I am mistaken) the seeds of various alpines on Ben Lawers, but so far as I can learn, none of these have been found on that hill. In hope that none of the plants thus sown would have established themselves, I did not mean to put the fact formally on record; but now that *Myosotis alpestris* has been actually recorded, it seems necessary to mention it. There is, however, one consolation for botanists, and that is that I believe all the seeds sown were of those plants which, from their showy flowers, are in common cultivation and hence that no obscure-flowered species, as grasses, sedges etc., if discovered hereafter in these localities, need be looked upon with suspicion." Only two appendices are required to this paragraph. It should be noted, first, that Snowdon too has suffered from a similar contamination of its native flora. As the result of the sowing, on a steep slope under the Parson's Nose, of the seeds of no less than two hundred and forty species of alpine plants, there were still in 1937-39 "thousands of seedlings" there of three alien species of *Primula*, "but no flowering plants" (cf. Blakelock, 1953, p. 181). And in the second place, there should be added to the list of plants sown in Caenlochan at least one other which has since been found there, the columbine *Aquilegia pyrenaica*. However desirable it may be to claim this beautiful plant as a native of British mountains, its cytology has betrayed it as an alien.

The last three of my selected areas are close enough together, all of them lying within what can strictly be called Glen Clova, to allow me to deal with them almost collectively. The stretch of cliff which encompasses Glen Doll and Glen Fee is so long—not to mention its height—that, even if I had surveyed every yard of it, it would still be impossible to attempt a description of the whole length. What I shall do, therefore, is simply to call in at the three points which I have found to be most interesting and describe very briefly the salient floristic features of each. And since once again most of the plants for which the district is noted have already been discussed, I shall concentrate attention on those mountain species which have not so far received more than a passing mention.

The classic locality for alpine milk-vetch (*Astragalus alpinus*) is in Glen Doll, where it was first discovered in 1831, but then, apparently, lost to sight again for more than half a century. Its favourite type of habitat in Britain is a grassy slope on shallow stony soil overlying basic rock; but in two of the three localities where I have seen it, it grows on the bare cliff as well. In Glen Doll it is accompanied by a rich selection of mountain plants, including boreal fleabane and the wintergreens, *Pyrola rotundifolia* and *Ramischia secunda*, rock speedwell, and above all, a magnificent display of mountain avens. Nearby, too, though not quite in the same association, are various mountain willows, including *Salix myrsinites*, and a relatively large and free-flowering colony of *Carex rupestris*. Hardly less interesting are the very steep slopes below, which are here and there covered with great tumbled boulders and scattered birches. On the flat top of some of these boulders is a thick layer of moss, and on at least one there grow from this layer, which has doubtless by now formed a fair depth of humus, fine colonies of dwarf cornel (*Chamaepericlymenum suecicum*, Pl. 13)

B. Perkins

Plate 13. Dwarf cornel, *Chamaepericlymenum suecicum.* Tolmount, Aberdeenshire

B. Perkins

Plate 14. Boreal fleabane, *Erigeron borealis*. Meall Garbh, Lawers range and lesser twayblade (*Listera cordata*). Among the heather are scattered colonies, one of them of some size, of *Linnaea borealis*—not strictly, perhaps, a mountain plant, but belonging to the same Scandinavian woodland association as the blue sow-thistle (*Cicerbita alpina*), and ascending in Scotland, like the sow-thistle, to an altitude of well over 2,000 feet. And finally, on the shingles of the Doll itself, and of one or two of its little tributaries, there is the beautiful species of lady's mantle, *Alchemilla conjuncta*, already mentioned in Chapter 8.

It is not by any means easy to understand what it is that determines the British distribution of alpine milk-vetch. The rock on which it grows in Glen Doll is said to be, and certainly appears to the layman, exactly the same hard quartzose schist as composes many neighbouring stretches of cliff on which the milk-vetch is not to be seen. On the other hand the yellow oxytropis (*O. campestris*), which inhabits one very short stretch above Glen Fee, would seem to be more easily explicable. At the point where it grows there is a narrow strip, from the top of the cliff to the bottom, of a rock which can be seen even from a distance to be lighter in colour than any other in the Glen; and while the oxytropis grows freely the whole way up this vertical strip, and even spreads some way down the scree at its foot, it does not stray at all on to the darker rock on either side. It was said long ago (by Professor Geikie, in *Transactions of the Botanical Society of Edinburgh, 12,* 159) that there is nothing peculiar about this rock either; but the visual evidence of one who is not a geologist still strongly suggests that there is. The case of the oxytropis is in fact one of those which shout loudest for an analysis of the rock upon which it grows, both in Glen Fee and near Loch Loch. In Glen Fee it is associated with no other flowering plants of comparable interest. There was a time, I believe, when *Woodsia ilvensis* could be seen growing with it, but the only station I now know for the *Woodsia* in the glen is some little distance away.

The last of all my selected areas lies also in Glen Fee, but this time it is not so much a vertical cliff as a steep bluff of rock between two long gullies. The most obvious floristic feature of this bluff is its display of mountain willows, among which can be numbered not only *Salix lanata* (Pl. XX, p. 177) which is its chief glory, but also *S. arbuscula, S. myrsinites* (Pl. XXI, p. 180), *S. herbacea* (Pl. 5, p. 64), *S. reticulata, S. lapponum* (Pl. XIX, p. 176), and a baffling series of hybrids. *Salix lanata* itself must be one of Britain's most beautiful plants, and when it grows, as it does here, with the other silver-silky willow, *S. lapponum*, the juxtaposition of the two heightens the beauty of each. It seems right that such a plant should be as rare as it is in Britain. Unfortunately the same cannot be said of the other special rarity of this bluff. The sedge *Carex norvegica*, which favours chiefly the damper patches at the very foot of the bluff, is one of those plants which a non-botanist would probably claim to have seen repeatedly on British hills; although enthusiasts for sedges might argue that in youth its compact black heads give it an air of distinction, by the time the fruits are ripe it has an undeniably common look. But the non-botanist would be wrong. Apart from this station and another, not far distant, over the watershed at the head of Glen Doll, its only certain British locality is on grassy slopes, at an altitude of about 3,000 feet, on the mountain near the head of Glen Lyon which I have already described as the best station for *Carex atrofusca*. There seems once again no good reason why the plant should not occur in other localities in Britain, but the fact remains that these are the only three areas where there is a reasonable probability of finding it.*

There is one other notable sedge in Glen Fee, namely *Carex grahami*, and a very puzzling plant it is. Its normal habitat seems always to be, as it is in Glen Fee itself, wet grassy slopes at altitudes between 2,000 and 3,000 feet. In such situations it has at one time or another been found on at least five different mountains, chiefly in Breadalbane; but though it is sometimes in considerable quantity over a few square yards, it seems invariably to grow in one very limited patch without outlying clumps or colonies. The belief that it is of hybrid origin has for the time being been abandoned, and it is now regarded instead as a separate mountain species, possibly endemic, most closely allied to the lowland sedge, *C. vesicaria*. That view may well be correct; but at the same time, in the few places where I have seen it in Scotland, it so evidently spreads vegetatively rather than by seed that until further investigations have been carried out the problem should probably not be regarded as finally solved. Moreover there is some reason to believe that the Glen Fee plant is not quite identical with that of the Breadalbanes. Admittedly in all its known stations *Carex grahami* is distinct enough from any other British sedge. But if there really are constant differences between the plant of one colony and that of another, there would seem to be taxonomic difficulties involved

* Since this was written *Carex norvegica* has been recorded also from the neighbourhood of Ben Lawers. See *Proc. B.S.B.I. 1*, 501.

M. C. F. Proctor

Plate XXI. A mountain willow, *Salix myrsinites.* Little Craigandal, Aberdeenshire

R. M. Adam

Plate XXII. Mountain bladder-fern, *Cystopteris montana.* Ben Lui here which do not apply, mercifully, to any of the other mountain sedges which I have mentioned.

(*d*) Postscript

So we come to the end of our tour of the mica-schist, and it leaves us, thanks to its tortuous length, with only two other notable plants of the district still to be discussed. *Athyrium flexile,* as it used to be called, is now thought to be a mere variety of the relatively widespread alpine lady-fern (*A. alpestre*); and there is no denying that on some mountains, chiefly in the Western Highlands, there are forms to be found intermediate between the two extremes. But since it is also undeniable that in some localities there are constant populations of a plant which is superficially widely different from the normal *A. alpestre,* a few words on the subject of the variety may not come amiss. Manton (1950) gives a description of the plant and discusses its claim to being treated as a distinct species. There is no difference in chromosome number between *A. flexile* and *A. alpestre,* nor for that matter between them and the common lady-fern, *A. filix-femina*; and further experimental taxonomic investigation is needed to decide the status of *A. flexile* and the forms intermediate between it and *A. alpestre.* This, Manton indicates, may be difficult, as the ferns seem to produce fertile fronds only sparingly in cultivation. Its best-known locality, unfortunately, is on one of the most inaccessible of the higher Scottish mountains, Ben Alder; and there, where it is apparently abundant, I have never gone to see it. I have however seen it, in what I can only think is comparable abundance, in one solitary corrie in the hills which border Rannoch Moor; and there it is not only liberally scattered over the cliffs, the gullies and the screes, but it is also so constant in appearance, with its dwarf stature, reflexed fronds and very short pinnae, as altogether to exclude, so far as I could see, anything approaching the normal *A. alpestre.* Since this particular corrie lies within the district treated in this chapter, and since a similar plant has evidently been found in the past in at least two of the glens of Angus, this seems the appropriate context in which to call attention to the existence of the variety. But I cannot help suspecting, for all that, that if its British distribution were properly known and charted, it too would prove, as I am sure normal *Athyrium alpestre* would, to be primarily calcifuge. Both the normal form and the variety, in fact, are probably instances of the type of plant which, being better suited on acid soils than on basic, are more widespread on the dull than on the rich mountains. As I have said in an earlier chapter, it is simply because such plants, thanks to leaching, can always find localities to suit them even on the richest hills, whereas the reverse process, by which calcicole plants occur on acid mountains, can be due only to geological abnormalities, that the flora of the basic mica-schist is so much richer than that of the acid granite.

Last but by no means least interesting of the plants of the mica-schist is the alpine coltsfoot (*Homogyne alpina*). Unfortunately—and this is the only reason why I have postponed the mention of it to the postcript—I have never myself succeeded in finding the *Homogyne* in Scotland, and know it only at first hand in the mountains of Austria. But one of the very few botanists who has seen it in its Scottish station, Mr. B. W. Ribbons, has very kindly sent us the following note: "The basal leaves of *Homogyne alpina* are kidney-shaped, dark-green and rather leathery. The stem is woolly, about a foot in height, often reddish-brown in colour, and bears a single head of florets which are tipped with purplish-red. It is not found in Scandinavia, but in the mountains of Central Europe and in the Pyrenees it is common up to altitudes of 10,500 feet, in boggy places in woods, pastures and heaths, especially where the soil is rich in humus. In the parish of Cortachy and Clova, in the County of Angus, *Homogyne alpina* grows on a wet hillside in association with such widespread mountain species as alpine meadow-rue, purple saxifrage, alpine lady's-mantle and mountain sorrel. The species is here in small quantity, and the number of plants which has been seen in flower is smaller still, and so if only a few botanists each collected a small 'piece', the plant would soon become extinct in Clova.

"George Don first found the alpine coltsfoot in Scotland [see p. 18], and in 1813 described it growing 'on rocks by the sides of rivulets in the high mountains of Clova, as on a rock called Garry-barns'. The species was apparently never seen again growing wild in Scotland until 1951, when Mr. A. Slack discovered a few square feet of it in what may be the same place as Don found it nearly 150 years ago. Like some of the other species which have been discovered or rediscovered in Scotland recently, the *Homogyne* grows in a situation which does not look 'botanically attractive', and so it is quite probable that no botanist since Don has been at the precise spot. Certainly it is doubtful whether any botanist since Don has ever tramped the Clova hills so systematically and so widely. Partly because the plant has never been seen since and partly because, over the years, Don has been given a reputation for unreliability, several eminent botanists have doubted the existence of *Homogyne alpina* in Scotland and it has never been admitted to the British flora. Very few British plants have the same kind of distribution outside Britain as *Homogyne* (one example is *Alchemilla conjuncta*) and this has encouraged the view that it is not a native British plant.

"Gardiner, the author of the *Flora of Forfarshire*, suggested that Don's plant might have been the fleabane (*Erigeron borealis*), and Druce thought that the leaves of dwarfed barren plants of the common coltsfoot (*Tussilago farfara*) had been mistaken for it. There are old herbarium specimens of the true plant from Scotland, at Kew and elsewhere; but these have been explained away by supposing that Don had plants from the Continent growing in his garden, and that he confused cultivated specimens with leaves from *Erigeron borealis* or *Tussilago farfara* actually collected on the hills.

"If Don did not really find the species in Angus, then either the plant was there but not discovered by him, or it has been introduced since by some unknown agency. Otherwise Don must have either deliberately planted it or else found it as he describes. In the absence of evidence to the contrary it seems most reasonable to accept his own account of the discovery of the plant. It is not possible to tell whether Don saw *Homogyne* in more than one spot, and there can be no proof that the existing patch is directly descended from the plants he found. But the species might now be removed from the list of 'Don's reputed discoveries'."

It is to be hoped that the deliberate lack of precision in this note about the plant's locality will stimulate further exploration of the less promising parts of the Clova district, and that other British stations may yet be found for an exceptionally interesting addition to the list of our mountain plants.

CHAPTER 10

IRELAND (J.R.)

In his admirable book *The Botanist in Ireland* Dr. R. Lloyd Praeger quotes the following eloquent sentences from an article by N. Colgan (in *Irish Naturalist*, 1900) entitled "Botanical Notes on the Galway and Mayo Highlands" :—

"It may sound like a paradox to say that the botanical survey of an Irish mountain region derives a peculiar zest from the very poverty of our flora in alpine species. Yet the assertion may be made with perfect truthfulness. That the rapture of discovery varies with the rarity of the object sought for, that the value of the thing attained is measured by the labour of attainment—these are time-honoured truisms in every system of proverbial philosophy; and their essential truth is daily borne in upon the mind of the botanist who devotes himself to the exploration of any of the mountain-groups of Ireland. The fans of the Alpine Club-moss, which he spurns with callous feet on the slopes of Snowdon, he half worships when they meet his longing eyes in the Wicklow or Kerry highlands; and so with many others of our alpine species—unconsidered trifles abroad, they become for him objects of enthusiasm at home."

These are the words, as Praeger says, of "one of the most critical of Irish botanists"; and Praeger's own book, in which special sections are devoted to the mountain flora of each range, goes to show how generally true they are. Two examples will suffice. The summit flora of Carrantual, the highest of Irish mountains, is typical of most of the rest. It consists, apparently, of only the following species;

Agrostis canina / *A. tenuis*	(bent)
Armeria maritima	thrift
Calluna vulgaris	heather

Deschampsia caespitosa	tufted hairgrass
D. flexuosa	wavy hairgrass
Festuca ovina	(fescue)
Galium hercynicum	heath bedstraw
Juncus bulbosus	(rushes)
J. squarrosus	
Luzula sylvatica	greater wood-rush
Lycopodium selago	fir club-moss
Potentilla erecta	tormentil
Rumex acetosa	sorrel
R. acetosella	sheep's sorrel
Saxifraga spathularis	St. Patrick's cabbage
S. stellaris	starry saxifrage
Sieglingia decumbens	heath grass
Thymus (presumably *T. drucei*)	wild thyme
Vaccinium myrtillus	bilberry

The botanical attractions of Ireland, great as they are, hardly extend to the mountain tops, and few visiting botanists are likely to tear themselves away from the fascinating flora of the Kerry lowlands for the sight of such a community. Nor will all botanists accept the verdict of Nathaniel Colgan that the very poverty of the flora adds zest to the exploration of an Irish mountain. For those who know North Wales or the Lake District, for instance, the knowledge that the parsley fern (*Cryptogramma crispa*) is in Ireland an extremely rare plant may well act as a discouragement rather than as a stimulus to the search. It is certainly an interesting fact that a plant so abundant in the mountains of England, Scotland and Wales should be so precariously established in Ireland that in one of its best localities, the Wicklow mountains, only five widely scattered plants of it have ever been seen, and in another, on the Garron Plateau, only six. But it is a fact which many botanists will regrettably prefer to accept on another's authority rather than laboriously to confirm on their own.

None the less there are at least two areas in Ireland which anybody with an interest in Britain's mountain flora will long to visit. It has been shown in earlier chapters how certain mountain plants tend to grow at much lower altitudes in the north and west of Scotland than they do in the central or eastern Highlands. This same phenomenon is at least equally noticeable in Ireland. Of seventeen mountain species which occur in both Donegal or Derry and in Kerry or Cork, the mean lower limit is, according to Praeger, only 550 feet in the north but nearly double that height in the south; and there is a similar decrease in the elevation of mountain plants towards the Atlantic seaboard in the west. The greatest number of mountain plants is to be found, as might be expected, in the north of the island. But they extend much more freely down the western side into Kerry than down the eastern into Wicklow, and it is no mere accident, but rather the result of a combination of climatic and geological factors, that both the areas to which I referred above are on or near the western coast.

We have already mentioned (pp. 82-3) what is undoubtedly the most remarkable area in Ireland for its mountain flora—the stretch of bare limestone on the coast of County Clare called the Burren Hills; and it is remarkable for four distinct reasons. In the first place, a number of rare and beautiful plants, such as mountain avens (*Dryas octopetala*), shrubby cinquefoil (*Potentilla fruticosa*) or spring gentian (*Gentiana verna* Pl. XXIII, p. 188 (×2)), are here in such profusion that the only other area in Britain that can rival the glamour of the Burren—and incidentally it is the only other area where these three species are all to be seen—is the valley of the Tees above Middleton in Teesdale (see Ch. 8(*a*)). In the second place, there are in the Burren, mixed with the arctic-alpine element represented by these three, a number of other plants, such as the little orchid *Neotinea intacta* or the maidenhair fern (*Adiantum capillus-veneris*), the European headquarters of which is around the Mediterranean. In the third place, this profusion of arctic-alpine plants grows on a tract of country which attains at its highest point only some 1,100 feet and which is notable for the mildness and dampness of its climate. No sight in Britain could be more surprising to a botanist, nor any more beautiful, than the masses of spring gentian which have spread off the limestone actually on to the sand-dune sward at Ballyvaughan; and every single one of the mountain species that are to be found in the Burren descends likewise to very few feet above sea-level. And finally, to add the last touch to the paradoxical character of the neighbourhood, there is the curious mingling of calcicole and calcifuge plants which is admirably illustrated, in a list cited by Praeger of the plants growing on a single flat slab of limestone, by the presence not only of mountain avens and the spring gentian, but also of such species as heather (*Calluna vulgaris*). This is indeed, as Praeger says, "a country of contradictions".

The fascination of the Burren has, however, been too often described to allow of a detailed account here, especially since it fell anyhow into the purview of Mr. Lousley's book in this series. Besides those plants already mentioned, others of the Burren which are particularly relevant to the present context are bearberry (*Arctostaphylos uva-ursi*), which is here very abundant; pyramidal bugle (*Ajuga pyramidalis*), which is to be found very locally, at altitudes of 100 feet or less, on either side of Galway Bay; blue moor-grass (*Sesleria caerulea*), one of the dominant constituents in the vegetation of the district; and, above all, two distinct forms of mossy saxifrage. The mossy saxifrages as a group have been in the past a cause of great difficulty and confusion to British botanists, and until 1950 as many as eleven species of the one group were recognised as British, of which no fewer than five were supposed to be confined to Ireland, while two more were at least more widespread there than elsewhere in the British Isles. In 1950, however, Professor D. A. Webb (Professor of Botany at Trinity College, Dublin) published a paper in *Proceedings of the Royal Irish Academy* in which confusion was at last reduced to order. It now appears that one of the two species growing in the Burren is indeed, as was always supposed, the widespread *S. hypnoides* itself, while the other, which was previously called *S. sternbergii* and supposed to be confined to four Irish counties, is merely one form of the polymorphic *S. rosacea,* which embraces also all the other forms that were previously regarded as endemic Irish species. There remains, even by Professor Webb's classification, one species of the group which is confined, in the British Isles at any rate, to one small portion of Ireland, namely the plant of the island of Arranmore which is now called *S. hartii*; but ironically enough the earlier botanists, despite their passion for increasing the group with new species, did not see fit to separate this particular plant from its relatives in Kerry or Clare. But for the rest, the task of the botanist is now greatly eased by the fact that he may legitimately class any mossy saxifrage which he finds in Ireland under either of the two highly variable species, *S. hypnoides* or *S. rosacea.*

The only district of Ireland which perhaps excels the Burren, and that not so much in its spectacular display as in its variety of rare mountain plants, is the Ben Bulben range of hills to the north and west of the town of Sligo. This, like the Burren, is an area of limestone; but it rises at its highest point to an altitude of nearly twice that of the highest point in the Burren, and no doubt chiefly for that reason harbours a considerable number of mountain plants which the Burren does not yield. Two of the three species which give to the Burren so much of its glamour, *Potentilla fruticosa* and *Gentiana verna*, are here absent; but their place is taken, on the cliffs and screes which surround the main massif, by an assortment of about twenty truly montane species. The length of cliff to be explored is very great, and some of the plants for which the district is notable are very far from being widespread there. But there are two relatively small areas which especially repay a visit. On or around the cliff called Annacoona, which overhangs the little hamlet of Gleniff, the special rarity of the district, Irish sandwort (*Arenaria ciliata* ssp. *hibernica*), has its headquarters; and it is accompanied in one of its stations there by a selection of mountain species including one other, in the shape of alpine saxifrage (*S. nivalis*), that is unknown elsewhere in Ireland. And further to the east, on the cliffs above Glenade, there is again a fair varity, of which perhaps the most distinguished is northern rock-cress (*Cardaminopsis petraea*), for which the only other Irish station is in the Galtees in Tipperary. It may well take a botanical visitor to the Ben Bulben range several days, if not weeks, before he has seen every one in the following list of the district's mountain plants:

Arenaria ciliata	Irish sandwort
Asplenium viride	green spleenwort
Cardaminopsis petraea	northern rock-cress
Carex bigelowii	(sedge)
Cochlearia alpina	mountain scurvy-grass
Draba incana	hoary whitlow-grass
Dryas octopetala	mountain avens
Epilobium alsinifolium	chickweed willow-herb
Oxyria digyna	mountain sorrel
Poa alpina	alpine meadow-grass
Polygonum viviparum	viviparous bistort
Polystichum lonchitis	holly fern
Salix herbacea	least willow
Saxifraga aizoides	yellow mountain saxifrage
S. hypnoides	mossy saxifrage
S. nivalis	alpine saxifrage
S. oppositifolia	purple saxifrage
Sedum rosea	rose-root

M. C. F. Proctor

Plate XXIII. Spring gentian, *Gentiana verna*. Burren, Co. Clare, Ireland

R. M. Adam

Plate XXIV. Yellow marsh saxifrage, *Saxifraga hirculus*. Pentland Hills

Silene acaulis	moss campion
Thalictrum alpinum	alpine meadow-rue.

But if he can spare a day for each of the two areas which (following as usual in Praeger's footsteps) I have particularly recommended, then he should at least have seen those plants in this list which the majority of British botanists would most wish to see.

In addition to its endemic sandwort, the Ben Bulben range provides an interesting and peculiar form of that most distinct of eyebrights, *Euphrasia salisburgensis*. On the Burren this eyebright is locally common, and very readily distinguishable from other species by its characteristic bushy growth, often coppery colour, and narrow, sharp-toothed leaves. Burren specimens differ slightly in general appearance from the form of *Euphrasia salisburgensis* common in the Swiss Alps, and H. W. Pugsley described them as a distinct variety *hibernica*. It seems, however, that Pugsley did not see the Ben Bulben representative; if he had, one feels, he would certainly have described it as different again, for the leaf-teeth are typically quite blunt, and the whole plant has a very different appearance. A careful study of the Irish eyebrights (*E. salisburgensis* is restricted to Ireland except for one unconfirmed record for the Yorkshire limestone) would prove very worthwhile; all that can be said at the moment is that the Ben Bulben form resembles some, but not all, of the plants called *Euphrasia lapponica* which occur in isolated populations in Northern Scandinavia.

If these two areas of limestone have a glamour which the higher Irish mountains cannot rival, yet few botanists visiting Ireland will readily omit from their itinerary either Kerry in the south or Donegal in the north; and in both counties they will find certain places where the mountain flora is not so contemptible as the list from the top of Carrantual might suggest. For our present purpose the richest areas in Macgillicuddy's Reeks are said to be the cliffs south of Loch Eagher, at the head of Cumloughra, and the coombs north of Lough Googh. But since the whole list of true mountain species of the district numbers barely a dozen, of which the most interesting are perhaps holly fern (*Polystichum lonchitis*) and alpine hair-grass (*Deschampsia alpina*), they are evidently rich rather in the luxuriance than in the variety of their vegetation. For my own part, what has struck me most about these hills, on one of the rare occasions when I have left the lowlands for the higher ground, was the very curious association which I found on the cliffs to the south of Lough Reagh. Two plants in particular were astonishingly abundant, each looking as much at home on the cliff as a plant could. One was the form of the saxifrage *S. rosacea* which used to be called *S. hirta*; the other was the little creeping willowherb, *Epilobium pedunculare,* which is a native of New Zealand. Since Praeger makes no mention at all of the presence of this little alien in Kerry, it has presumably only recently arrived there. It is certainly spreading elsewhere in the British Isles at a remarkable speed, and especially in mountainous districts. I do not remember ever having seen it before the last war, though no doubt it was already beginning to establish itself here and there. Since the war I have seen it in at least one locality in each of the four countries—on the moors near Trawsfynydd, for instance, in Wales; on grassy banks both in Glen Lyon in Perthshire and in the northern tip of Skye; and, above all, at intervals all the way up Helvellyn from Glenridding. In none of these stations, however, is it so difficult to realise that it is beyond doubt an alien as it is on the Kerry cliff. Presumably it is its possession of long-plumed seeds—a characteristic which it shares with the other willowherbs—that enables it to spread at one step not mere inches but many miles; and since it is evidently much addicted to damp stony places, it has successfully made for itself a place in the mountain flora of Britain to which no other alien can aspire.

As Praeger says of the Reeks, "the flora increases rapidly in variety as one descends the hills"; nor does one have to descend far before one meets with a number of the plants for which Kerry is especially famed. St. Patrick's cabbage (*Saxifraga spathularis*) actually ascends, as we have already seen, to the very top of Carrantual, while its relative, kidney saxifrage (*S. hirsuta*), reaches an elevation of some 3,000 feet, and even the large-flowered Irish butterwort (*Pinguicula grandiflora*) climbs up to 2,800. Kerry and Donegal are in fact in this respect sharply contrasted, that whereas in Kerry certain plants of the "Lusitanian" or Pyrenean group attain an altitude which brings them strictly within the compass of this book, in Donegal several of the "Highland" group descend right down to sea-level. Hoary whitlow-grass for instance, ranges in Donegal from sea-level to only about 250 feet, purple saxifrage from sea-level to 1,850 feet and rose-root from sea-level to 1,900 feet; and alpine meadow-rue (*Thalictrum alpinum*), alpine saw-wort (*Saussurea alpina*) and mountain sorrel all descend here, as they do also in north and west Scotland, below the

B. Perkins

Plate 15. Alpine cinquefoil, *Potentilla crantzii.* Ben Lawers

B. Perkins

Plate 16. Rock speedwell. *Veronica fruticans.* Meall nan Tarmachan, Perthshire

1,000 foot contour. The mountain plant with the highest average elevation in Donegal is holly fern, with a range of 1,470 to 1,850 feet.

The most interesting mountain in Donegal for the botanist (if it can be called a mountain) is undoubtedly Slieve League, which rises in one vast precipice from the Atlantic, on the north side of Donegal Bay, to a height of just under 2,000 feet. Its northern face, which is only slightly less sheer than its southern, is the home of a larger selection of mountain plants than any other area of comparable size in Ireland can boast, and it is only because that selection comprises nothing of comparable interest with Irish sandwort or spring gentian that I have not classed Slieve League, with Ben Bulben and the Burren, as a place that must be visited. Indeed nothing could give a better picture of the mountain flora of Ireland than the list which Praeger cites of the "alpine" plants which grow on or near this one cliff. It runs thus:

Arctostaphylos uva-ursi	bearberry
Asplenium viride	green spleenwort
Carex bigelowii	(sedge)
Dryas octopetala	mountain avens
Hieracium anglicum *H. iricum*	(hawkweeds)
Juniperus communis ssp. *nana*	dwarf juniper
Lycopodium alpinum	alpine club-moss
Oxyria digyna	mountain sorrel
Polygonum viviparum	viviparous bistort
Polystichum lonchitis	holly fern
Salix herbacea	least willow
Saussurea alpina	alpine saw-wort
Saxifraga aizoides	yellow mountain saxifrage
S. oppositifolia	purple saxifrage
S. stellaris	starry saxifrage
Sedum rosea	rose-root
Selaginella selaginoides	lesser club-moss
Thalictrum alpinum	alpine meadow-rue
Vaccinium vitis-idaea	cowberry

It would be hard to find, in any other comparable area of the British

Isles, so large a selection of the commoner mountain plants without at least one real rarity thrown in.

If I seem to have done scant justice to the mountains of the eastern half of Ireland, that is chiefly, I admit, because I have never explored them; but that fact in itself is not entirely irrelevant. The best that the Mourne Mountains can produce in the way of a mountain flora consists only of starry saxifrage, rose-root, alpine saw-wort, least willow and alpine club-moss; while the mountains of Wicklow, despite their extent, can add to this uninspiring list no mountain plant of greater interest than mossy saxifrage. Nathaniel Colgan's verdict on the Galway and Mayo Highlands was quoted at the beginning of the chapter. His assessment of that part of the Wicklow Mountains which lies in County Dublin (an assessment which is again quoted by Praeger) is hardly less eloquent. The flora of the ground lying between the 2,000 and the 2,450 foot contours "differs to only a trifling extent from the flora of an Irish lowland bog." Indeed, much the most interesting collection of mountain plants in the eastern half of Ireland is to be found, usually at quite low altitudes, in the Tertiary basalt region of Antrim and Derry; and once again we see the importance of the influence of calcareous rock. The basalt near the coast, most famous scenically at the Giant's Causeway, is the home of a number of northern plants, some of them, such as rose-root, truly arctic-alpine. The finest display is to be seen on Benevenagh, where the basaltic cliffs rise to little over 1,200 feet, but where moss campion, purple saxifrage, and mountain avens all occur. Inland too the Garron plateau, a bare, deserted expanse of moorland and bog, provides localities for the rare saxifrage *S. hirculus,* and the only known Irish stations for two Scottish mountain sedges, *Carex pauciflora* and *C. paupercula.* Again it seems reasonable to suppose that the presence of these rare northern species in this area has something to do with the availability of rather base-rich drainage water from the volcanic rocks.

From this very cursory survey, therefore, three main facts would seem to emerge. The first can once again be stated, and very briefly too, in the words of Praeger: "the 'Highland' plants of Watson as a group do not in Ireland fit well with his definition, 'species chiefly seen about the mountains'." And if that fact is itself of little immediate use to the visiting field botanist, then its two corollaries may prove more practical. For a proper appreciation of the "Highland" flora of Ireland, the first essential is to concentrate on the extreme west and north-west, where the Atlantic climate produces its paradoxical effect. And even when the field has been thus drastically narrowed, there still remains a golden rule which seldom fails: go first to those areas where the acid rocks, which predominate down the western coast of Ireland, give place for an appreciable stretch to the basic. At least the simultaneous observance of these two rules will lead straight to the areas on which I have particularly dwelt; and it may well be that an even more thorough exploration of the other regions thus indicated would yet yield some unexpected and exciting results.

CONCLUSION (M.W.)

Our brief and inadequate tour of British mountains has, I hope, excited the interest of the reader to see for himself something at least of the rich flora they possess, and, for those already with some knowledge of where the interesting mountain plants grow, it has not, perhaps, been entirely unrewarding. I make no apology for re-emphasising here the heightened responsibility that a knowledge of British mountain plants should bring, in the direction of preservation for future generations of the interest and beauty of this mountain flora. The present century has seen a very welcome growth of interest in the conservation of nature, and since the second world war there have been created the first nature reserves in the British mountains—the Ben Lawers reserve owned by the National Trust, and the great reserves of the Cairngorms and Beinn Eighe, and the very recently-declared reserve of Cwm Idwal in Snowdonia, all owned and administered as National Nature Reserves by the Nature Conservancy. In all these reserves reasonable access is allowed, but *collecting* of plants or animals is restricted. The collector's urge is strong within many of us; but, in the case of mountain plants in particular, we must sublimate it into more useful and less selfish directions. Photography, particularly colour photography, is one obvious way of doing this: mountain plants—as we hope some of the pictures in this book testify—make excellent subjects for the photographer's art, and a most valuable record of how the plant grows in its natural home.

For those who would like to know in what directions the interested and energetic amateur can make his hobby serve scientific ends, the study of the mountain flora offers quite exceptional scope. As we have had several occasions to remark throughout the book, considerable areas of the mountainous parts of Scotland have been so little visited by botanists that it is still possible to make major discoveries of distinct mountain plants not previously known in Britain. Moreover, even if every day on an unknown Scottish mountain does not produce its

Diapensia, the information on distribution of the British mountain flora as a whole is woefully inadequate in many areas, and simple records of the plants growing on a particular mountain can be really useful. This is especially so at the present time, when the 5-years' scheme to map the distribution of British flowering plants and ferns, inaugurated in April, 1954, by the Botanical Society of the British Isles, is in urgent need of such basic information, particularly for areas of mountain and moorland off the botanical 'beaten track'. So, if you want to contribute some really useful data, tell us what grows in particular places on, for example, the many botanically unknown mountainous areas of Ross-shire. By all means, of course, visit Ben Lawers and Clova too; but such well-known places should only provide your initiation into British mountain botany, and the investigation of new territory the real specialist delight! It is a well-known fact of human experience that the discovery made for oneself makes a far more lasting impression than that provided by a 'conducted tour'; and this is nowhere more true than in botanising on mountains.

As to more serious study, there is no shortage of problems in connexion with the mountain flora which are well within the scope of the keen amateur. It is probably true to say that there is not a single rare or local mountain plant in the British flora about which we have anything approaching adequate information to form a basis for an explanation of its restricted occurrence. Of course, as we have often exemplified in this book, we can sometimes point to *some* factors of importance, such as the presence of lime in the soil, or the intolerance of shade; but even where this amount of analysis is possible, we still do not know how this factor operates in terms of restricted growth or mortality of plants or seedlings. And the whole question of seed production and seedling establishment in mountain plants is an almost virgin field for investigation. Again, the study of the relationships between mountain plants and animals (as, for example, in the case of insects and the plants they feed on and pollinate) would provide many fascinating possibilities for the field naturalist's attention. Finally, 'critical' groups, needing experimental taxonomic study, are especially well represented in the mountain flora; in addition to the hawkweeds (*Hieracium*) and the eyebrights (*Euphrasia*), which we have already discussed, one might mention scurvy-grass (*Cochlearia*) and pearlworts (*Sagina*) as further examples which would yield much useful information to patient study.

Rich and varied though it is, the flora of the British mountains is really a poor show when compared with those of other parts of Continental Europe. Logically, therefore, a visit to the Scandinavian mountains and the Alps—in that order—should follow some acquaintance with British mountains. Each step is a revelation of a whole further dimension, as it were, of mountain vegetation—the Scandinavian mountains showing us roughly what the richest parts of the Scottish hills might offer if sheep-grazing had never become the standard practice, and the Alps adding an almost overwhelming further variety to the mountain flora. There is, I suppose, a danger that one might lose one's heart to the Alpine flora, and turn one's back for ever on the dowdy mist- and rain-washed hills of Highland Britain. I can only say that this has not happened to me; spring gentian in Teesdale or the drooping saxifrage in Glencoe still provide a thrill quite undiminished by the fact that I have seen spring gentians by the thousand in the Alps or walked over and ignored the drooping saxifrage in Scandinavia. Familiarity with the British mountain flora need never breed contempt.

APPENDIX I

MOUNTAIN PLANTS

Agropyron donianum	'Don's twitch'
**Ajuga pyramidalis*	pyramidal bugle
Alchemilla alpina	alpine lady's-mantle
* ,, *conjuncta*	,,
,, *filicaulis*	
,, *glomerulans*	(lady's mantle)
,, *wichurae*	
Alopecurus alpinus	alpine foxtail
Antennaria dioica	cat's-foot
Arabis alpina	alpine rock-cress
Arctous alpina	black bearberry
**Arenaria ciliata* ssp. *hibernica*	Irish sandwort
**Arenaria gothica*	Fries' sandwort
* ,, *norvegica*	Norwegian sandwort
Armeria maritima	thrift
Artemisia norvegica	
Asplenium viride	green spleenwort
Astragalus alpinus	alpine milk-vetch
Athyrium alpestre (incl. *A. flexile*)	alpine lady-fern
Bartsia alpina	
Betula nana	dwarf birch
Cardaminopsis petraea	northern rock-cress
Carex aquatilis	
,, *atrata*	
,, *atrofusca*	
,, *bigelowii*	
,, *capillaris*	
(*,, *davalliana*)	
,, *grahami*	(sedges)
,, *x helvola*	
,, *lachenalii*	
,, *microglochin*	
,, *norvegica*	
,, *pauciflora*	
,, *paupercula*	
,, *rariflora*	
Carex rupestris	
,, *saxatilis*	(sedges)
,, *vaginata*	
Cerastium alpinum	alpine mouse-ear chickweed
,, *cerastoides*	(mouse-ear chick-weed)
,, *edmonstonii*	arctic mouse-ear chickweed
Chamaepericlymenum suecicum	dwarf cornel
Cherleria sedoides	mossy cyphel
Cicerbita alpina	blue sow-thistle
Cochlearia alpina	mountain scurvy-grass
,, *micacea*	Scottish scurvy-grass
Cryptogramma crispa	parsley fern
Cystopteris montana	mountain bladder-fern
Deschampsia alpina	
Diapensia lapponica	
Draba incana	hoary whitlow-grass
,, *rupestris*	rock whitlow-grass
Dryas octopetala	mountain avens
Empetrum hermaphroditum	(crowberry)
Epilobium alsinifolium	chickweed willow-herb
,, *anagallidifolium*	alpine willow-herb
**Equisetum variegatum*	variegated horsetail
Erigeron borealis	boreal fleabane
Euphrasia cambrica	
,, *frigida*	
,, *rivularis*	(eyebrights)
* ,, *salisburgensis*	
,, *scotica*	

Festuca vivipara	viviparous fescue
Galium boreale	northern bedstraw
,, pumilum	slender bedstraw
Gentiana nivalis	snow gentian
* ,, verna	spring gentian
Gnaphalium norvegicum	highland cudweed
,, supinum	dwarf cudweed
Hieracium spp.	(hawkweed)
Homogyne alpina	alpine coltsfoot
Juncus alpinus agg.	alpine rush
,, biglumis	two-flowered rush
,, castaneus	chestnut rush
,, trifidus	three-leaved rush
,, triglumis	three-flowered rush
Kobresia simpliciuscula	
Koenigia islandica	
*Linnaea borealis	
Lloydia serotina	
Loiseleuria procumbens	mountain azalea
Luzula arcuata	curved woodrush
,, spicata	spiked woodrush
Lycopodium alpinum	alpine clubmoss
,, annotinum	interrupted clubmoss
,, clavatum	stag's-horn moss
,, selago	fir clubmoss
Minuartia rubella	alpine sandwort
* ,, stricta	Teesdale sandwort
* ,, verna	vernal sandwort
Myosotis alpestris	alpine forget-me-not
Oxyria digyna	mountain sorrel
Oxytropis campestris	yellow oxytropis
* ,, halleri	purple oxytropis
Phleum commutatum	
Phyllodoce coerulea	
(*Pinguicula alpina	alpine butterwort)
,, grandiflora	Irish butterwort
Plantago maritima	sea plantain
Poa alpina	alpine meadow-grass
,, balfourii	
,, flexuosa	
,, glauca	
,, x jemtlandica	
Polygonum viviparum	viviparous bistort
Polystichum lonchitis	holly fern
Potentilla crantzii	alpine cinquefoil
* ,, fruticosa	shrubby cinquefoil
Rhinanthus borealis agg.	(yellow rattle)
(Rubus arcticus)	
,, chamaemorus	cloudberry
,, saxatilis	stone bramble

Sagina intermedia	lesser alpine pearlwort
,, saginoides	alpine pearlwort
Salix arbuscula	(willow)
,, herbacea	least willow
,, lanata	woolly willow
,, lapponum	(willow)
,, myrsinites	(willow)
,, reticulata	reticulate willow
Saussurea alpina	alpine saw-wort
Saxifraga aizoides	yellow mountain saxifrage
,, caespitosa	tufted saxifrage
,, cernua	drooping saxifrage
* ,, hirculus	yellow marsh saxifrage
,, hirsuta	kidney saxifrage
,, hypnoides	mossy saxifrage
,, nivalis	alpine saxifrage
,, oppositifolia	purple saxifrage
,, rivularis	brook saxifrage
,, rosacea	
,, spathularis	St. Patrick's cabbage
,, stellaris	starry saxifrage
Sedum rosea	rose-root
,, villosum	hairy stonecrop
Selaginella selaginoides	lesser clubmoss
Sesleria caerulea	blue moor-grass
Sibbaldia procumbens	
Silene acaulis	moss campion
* ,, maritima	sea campion
Taraxacum spp.	(dandelion)
Thalictrum alpinum	alpine meadow-rue
*Thlaspi alpestre	alpine penny-cress
Tofieldia palustris	Scottish asphodel
(*Trichophorum alpinum	alpine cotton-grass)
Trientalis europaea	chickweed wintergreen
Vaccinium uliginosum	bog whortleberry
Veronica alpina	alpine speedwell
,, fruticans	rock speedwell
Viola lutea	mountain pansy
* ,, rupestris	Teesdale violet
Viscaria alpina	red alpine catchfly
Woodsia alpina	
,, ilvensis	

PLANTS NOT UNCOMMONLY OCCURRING ABOVE 2,000 FEET ON MOUNTAINS IN BRITISH ISLES

Achillea millefolium	yarrow
Adoxa moschatellina	moschatel
Agrostis canina	brown bent-grass
„ *tenuis*	common bent-grass
Ajuga reptans	bugle
Alchemilla glabra	(lady's mantle)
Anemone nemorosa	wood anemone
Angelica sylvestris	wild angelica
Anthoxanthum odoratum	sweet vernal-grass
Anthyllis vulneraria	kidney-vetch
Arabidopsis thaliana	thale cress
Arctostaphylos uva-ursi	bearberry
Bellis perennis	daisy
Blechnum spicant	hard fern
Briza media	quaking grass
Calluna vulgaris	heather, ling
Caltha palustris	kingcup
Campanula rotundifolia	harebell (in Scotland, bluebell)
Cardamine flexuosa	wood bitter-cress
„ *hirsuta*	hairy bitter-cress
„ *pratensis*	cuckoo flower
Carex binervis (incl. *C. sadleri*)	(sedges)
„ *curta*	
„ *dioica*	
„ *echinata*	
„ *flacca*	
„ *nigra*	
„ *ovalis*	
„ *panicea*	
„ *pilulifera*	
„ *pulicaris*	
Cerastium vulgatum	common mouse-ear chickweed
Chamaenerion angustifolium	rosebay willow-herb
Chrysosplenium oppositifolium	opposite-leaved golden saxifrage
Coeloglossum viride	frog orchid
Conopodium denudatum	pignut
Crepis paludosa	marsh hawk's-beard
Cystopteris fragilis	brittle bladder-fern
Deschampsia caespitosa	tufted hair-grass
„ *flexuosa*	wavy hair-grass
Digitalis purpurea	foxglove
Drosera anglica	great sundew
„ *rotundifolia*	sundew

Dryopteris austriaca	broad buckler-fern
„ *filix-mas* agg.	male fern
Eleocharis pauciflora	few-flowered spike-rush
Empetrum nigrum	crowberry
Epilobium montanum	broad-leaved willow-herb
„ *obscurum*	(willow-herb)
„ *palustre*	marsh willow-herb
Equisetum arvense	common horsetail
Erica cinerea	bell-heather
„ *tetralix*	cross-leaved heath
Eriophorum angustifolium	(cotton-grass)
„ *vaginatum*	
Erophila verna agg.	whitlow grass
Euphrasia micrantha	(eyebright)
Festuca ovina	(fescue)
„ *rubra*	
Filipendula ulmaria	meadow-sweet
Galium hercynicum	heath bedstraw
„ *palustre*	marsh bedstraw
Geranium sylvaticum	wood cranesbill
Geum rivale	water avens
Heracleum sphondylium	cow parsnip
Hieracium spp.	(hawkweed)
Holcus lanatus	Yorkshire fog
Juncus acutiflorus	sharp-flowered rush
„ *articulatus*	jointed rush
„ *bufonius*	toad rush
Juncus bulbosus	bulbous rush
„ *conglomeratus*	(rush)
„ *effusus*	soft rush
Leontodon autumnalis	autumnal hawkbit
Linum catharticum	purging flax
Listera cordata	lesser twayblade
Lotus corniculatus	birdsfoot-trefoil
Luzula multiflora	many-headed wood-rush
„ *sylvatica*	greater woodrush
Lychnis flos-cuculi	ragged robin
Melampyrum pratense	common cow-wheat
Melandrium rubrum	red campion
Molinia caerulea	purple moor-grass
Montia fontana agg.	blinks
Myosotis secunda	water forget-me-not
Nardus stricta	mat-grass
Narthecium ossifragum	bog asphodel

Oxalis acetosella	wood-sorrel
Oxycoccus palustris (incl. *O. microcarpus*)	cranberry
Parnassia palustris	grass of Parnassus
Pedicularis palustris	red rattle
,, *sylvatica*	lousewort
Pinguicula vulgaris	common butterwort
Poa annua	
,, *nemoralis*	(meadow-grass)
,, *pratensis*	
,, *trivialis*	
Polygala serpyllifolia	(milkwort)
,, *vulgaris*	
Polypodium vulgare	polypody
Potentilla erecta	common tormentil
Prunella vulgaris	self-heal
Ranunculus acris	meadow buttercup
Rununculus flammula	lesser spearwort
,, *repens*	creeping buttercup
Rhinanthus minor agg.	(yellow rattle)
Rumex acetosa	sorrel
,, *acetosella*	sheep's sorrel
Sagina procumbens	procumbent pearl-wort
Salix repens	creeping willow
Sedum anglicum	English stonecrop
Sieglingia decumbens	heath grass
Solidago virgaurea	golden-rod
Stellaria alsine	bog stitchwort
Succisa pratensis	devil's-bit scabious
Taraxacum spp.	(dandelion)
Thelypteris dryopteris	oak fern
,, *oreopteris*	mountain fern
,, *phegopteris*	beech fern
Thymus drucei	(wild thyme)
Trichophorum caespitosum	'deer-grass'
Trifolium repens	white clover
Trollius europaeus	globe-flower
Tussilago farfara	coltsfoot
Urtica dioica	stinging nettle
Vaccinium myrtillus	bilberry (in Scotland, blaeberry)
,, *vitis-idaea*	cowberry
Veronica chamaedrys	germander speedwell
,, *officinalis*	common speedwell
,, *serpyllifolia* (incl. ssp. *humifusa*)	thyme-leaved speedwell
Viola palustris	marsh violet
,, *riviniana*	common violet

An English name enclosed in brackets applies to a group of species, not merely to the one in question.

A number of other mountain plants have from time to time been recorded in the British Isles (see, for instance, pp. 18 and 177) but have been omitted here, as, in the opinion of the authors, the records need confirmation.

* Indicates a 'mountain' species with few or no British localities above 2,000 feet. Species in brackets were recorded in the nineteenth century but are now apparently extinct in Britain.

APPENDIX II

(*a*) KEY TO THE LADY'S MANTLES (ALCHEMILLA) OCCURRING ABOVE 2,000 FT. ON BRITISH MOUNTAINS (ON LEAF-CHARACTERS ONLY)

1 Leaves with 5 or 7 separate digitate leaflets, green above and silvery beneath. Dwarf plant, locally common. *A. alpina*

A. conjuncta, similar, but with leaflets joined at the base, does not grow above 2,000 ft. in any British locality.

Leaves lobed, green on both sides. *A. vulgaris* agg. 2

2 Leaves with some hairs *spreading at right angles*, either on petiole, or upper surface, or both. Common on Scottish mountains; also recorded in N. England, N. Wales and Ireland. *A. filicaulis*

Leaves with only more or less appressed hairs, or apparently glabrous. 3

3 Leaves usually with some hairs on *upper* surface; leaf when fully-developed wavy in outline, often rather yellow-green and 'oily' in appearance. High Scottish mountains and Upper Teesdale only. *A. glomerulans*

Leaves entirely glabrous on upper surface; leaf when fully-developed more or less in one plane. 4

4 Leaf-outline circular, with basal lobes meeting or even partly overlapping; teeth very neat and sharp; a narrow toothless V-shaped incision between the leaf-lobes; leaf-texture thin and papery. On wet basic rock on Scottish mountains and in N. England; not recorded from Wales or Ireland. *A. wichurae*

Leaf-outline various, but not often circular, usually with widely separated basal lobes; teeth much coarser and less regular; little or no V-shaped incision between lobes; leaf-texture coarser. Much the commonest member of the *vulgaris* agg. on mountains. *A. glabra*

If flowering specimens are available, it is easier to be certain of *A. glomerulans*, which has obvious loosely appressed silky hairs right up the stalk into the branches of the inflorescence. *A. filicaulis* has *spreading* hairs only in the lower part of the stalk, and *A. glabra* and *A. wichurae* have appressed hairs only in the same regions.

(*b*) KEY TO THE "ALPINE" HAWKWEEDS (HIERACIUM) IN BRITAIN

On British mountains the great majority of the yellow "dandelion-like" *Compositae* are species of hawkweeds (*Hieracium*). Since, however, four other yellow Composites do occur more or less frequently above 2,000 feet, it may be as well to eliminate these at the outset.

First is the coltsfoot (*Tussilago farfara*), a familiar plant of lowland Britain. When it is in flower the coltsfoot is unmistakable: the solitary yellow flower-heads are borne on scaly, woolly stalks, which are often reddish or purplish in colour, and they appear long before the large, roundish leaves. In the lowlands the coltsfoot flowers in February or March, but on mountains, where it may be snow-covered for several months, it can be seen in flower in June or even later.

Next there is the dandelion itself (*Taraxacum*), which, like *Hieracium*, is an apomictic genus containing many microspecies, but which, unlike *Hieracium*, has so far proved too difficult a problem for British botanists to tackle (see p. 75). All dandelions can be easily recognised by their smooth, hollow flower-stems, which contain copious milky juice. The mountain dandelions usually have short stalks and grow commonly in wet flushes.

The third of the four, which may often be found in similar places on mountains, is the autumnal hawkbit (*Leontodon autumnalis*). This differs from the dandelion in having a more wiry flower-stalk, with several scattered bracts below the flower-head, and in being usually branched (though dwarf mountain plants sometimes bear solitary flower-heads). From all *Hieracia* it may most readily be distinguished by the *numerous* bracts below its flower-heads.

And finally there is the marsh hawk's-beard (*Crepis paludosa*). That this should so often be mistaken for a hawkweed is not surprising, for technically it is very close to being one. It is typically a tall plant of marshy ground, with several amplexicaul stem-leaves, and it is best distinguished from similar-looking hawkweeds by its *glabrous* stem and leaves, and by its involucral bracts, which are clearly in *two* rows only, the outer bracts being only about half as long as the inner.

Any botanist who is familiar with these four may be pretty confident that any other yellow-flowered Composite that he finds on a British mountain is a hawkweed. The following key is concerned only with the more widespread members of the two sections *Alpina* and *Subalpina*. These two sections as a whole can be fairly easily distinguished from all other sections in Britain except the *Cerinthoidea* by their combination of small stature, few stem-leaves, single- or few-headed inflorescence, large, dark and generally long-pilose involucres, and showy flower-heads. From the *Cerinthoidea* they can usually (but in the case of the *Subalpina* not always) be distinguished by the presence on the leaves, especially on the margins near the base, of a certain number of small yellowish stalked glands. As with all critical or difficult genera, so particularly with the hawkweeds, of which many in Britain are as yet un-named, there is ultimately no substitute for familiarity with the plants in the field. If it is used as it is meant to be, the key may perhaps help towards the identification of some of the mountain species. If it fails, that will probably be because it is excessively simplified; but it will be as well to remember the possibility that the species in question has not yet been described.

In the section *Alpina* all except six species in Britain—*H. alpinum, H. holosericeum, H. eximium, H. calenduliflorum, H. gracilentum, H. globosiflorum*—appear to be endemic; in the section *Subalpina* all except *H. molybdochroum, H. gracilifolium* and *H. submurorum*. Of the nine exceptions five—*H. alpinum, H. holosericeum, H. eximium, H. calenduliflorum* and *H. gracilentum*—are found outside Scotland in both Scandinavia and the mountains of Central Europe; the remaining four in Scandinavia only.

Key to British Hieracia of Sections Alpina and Subalpina

This key is not intended to be used on isolated individual specimens, which anyhow often defy determination even by an expert, but only on typical representatives of a colony *in a fresh state.* The colour of leaves, and even more of styles, often changes completely in the process of pressing.

For the sake of brevity I have made use in the key of certain technical terms, as follows:
ligule = the flattened part of the floret, above the tube
phyllary = involucral bract

The hair-clothing of hawkweeds is of three kinds:
floccose = minute, white, star-shaped hairs
glandular = stalked glands, like tiny knitting-needles
pilose = simple whitish hairs, in the *Alpina* and *Subalpina* usually dark-based, long and silky (epilose = without any hairs of this last kind).

Majority of plants in a colony with solitary heads — ALPINA
Majority of plants in a colony with 2 or more heads — SUBALPINA

Section ALPINA

1 Styles (in fresh, not pressed, plants) pure yellow — 2
Styles discoloured to black — 6

2 Majority of radical leaves 2 to 3 times as long as broad — 3
Majority of radical leaves more than 3 times as long as broad — 5

3 Leaves irregularly incise-dentate — *H. hanburyi*
Leaves subentire or with few denticulations — 4

4 Upper side of leaves (especially stem-leaves) densely pilose with long stiff white hairs — *H. alpinum*
Upper side of leaves only sparsely pilose or glabrescent — *H. globosiflorum*

5 Leaves all subentire — *H. holosericeum*
Some leaves sharply and irregularly dentate — *H. eximium* var. *tenellum*

6 Majority of radical leaves less than 3 times as long as broad — 7
Majority of radical leaves more than 3 times as long as broad — 10

7 Leaves deep blue-green, sometimes spotted with purplish-brown — *H. calenduliflorum*
Leaves light- or yellowish-green — 8

8 Ligules strongly pilose on the back as well as the tips — *H. subgracilentipes*
Ligules slightly pilose at the tips only — 9

9 Radical leaves subentire or with few small teeth, styles dirty yellow to brown — *H. gracilentum*
Radical leaves sinuate-dentate or coarsely toothed, styles black — *H. pseudo-curvatum*

10 Radical leaves yellowish-green, densely pilose — *H. eximium*
Radical leaves dark blue-green, sparsely pilose or glabrescent — *H. backhousei*

For the sake of simplicity I have omitted from the key to the Section ALPINA (*a*) the four species confined to the Cairngorms (*b*) certain described varieties of *H. globosiflorum* and *H. hanburyi*. Of the Cairngorm species, however, two are easily recognised, *H. grovesii* by its almost epilose but densely floccose leaves, *H. macrocarpum* by the fact that it often bears 2 heads, held on long sub-equal peduncles so that the whole stem forms a Y.

Section Subalpina

1 Pilose hairs on involucre far outnumbering glandular 2
Glandular hairs on involucre roughly equalling or far outnumbering pilose 6
2 Radical leaves deep bluish-green, paler beneath *H. dasythrix*
Radical leaves light green, yellowish or glaucescent 3
3 Leaves lingulate, rough with stiff hairs, subentire or with few distant small teeth or denticulations *H. lingulatum*
Radical leaves sharply and coarsely toothed 4
4 Leaves very irregularly toothed *H. cremnanthes*
Leaves more or less regularly toothed 5
5 Some radical leaves obovate; ray florets few and long (i.e. head radiant) *H. callistophyllum*
Some radical leaves elliptic-lanceolate, but never obovate; ray florets numerous and normal *H. marshallii*
6 Glandular hairs on involucre far outnumbering pilose 7
Glandular and pilose hairs on involucre roughly equal in number 9
7 Involucres epilose; leaves often richly spotted with purplish-brown *H. gracilifolium*
Some pilose hairs present on involucre; leaves unspotted 8
8 Heads turbinate (top-shaped) or obconic *H. hyparcticoides*
Heads rounded, slightly narrowed below *H. submurorum*
9 Styles yellow *H. senescens*
Styles dark 10
10 Heads rounded below, leaves oblong to lanceolate
Heads not rounded, leaves elliptic-lanceolate
11 Leaves subentire or denticulate; phyllaries long, narrow and acute *H. pseudanglicum*
Leaves sinuate-dentate, often deeply; phyllaries broad and obtuse *H. sinuans*
12 Heads cylindrical, abruptly narrowed below; phyllaries black, ligules strongly pilose-tipped *H. centripetale*
Heads turbinate, subtended by bracteoles; phyllaries olive-green; ligules glabrous-tipped *H. isabellae*

In this section I have omitted *H. molybdochroum*, which is known in Britain only from Cairntoul in the Cairngorms, and *H. cumbriense*, which is confined to the English Lake District and is the only species of the section known there.

(c) KEY TO THE MOUNTAIN WILLOWS (SALIX) IN BRITAIN

The following key, which is based on Dr. Warburg's in the *Flora of the British Isles,* is intended for use on plants (of either sex) with mature leaves. Any other key would necessarily be more complicated; and in any case the majority of the species concerned produce their leaves before or with the catkins. By 'mountain willows' I mean only those prostrate or small shrubs which are especially characteristic of the mountains; but I have included *Salix repens* because it too is a small shrub which is not infrequently found above 2,000 feet. Other willows occurring above 2,000 feet on British mountains are *S. aurita, S. caprea, S. nigricans* and *S. phylicifolia.* Numerous hybrids have been reported, many of them on reliable authority, both between one mountain species and another, and between mountain and lowland species. I have included no such hybrids in the key, for their determination is always difficult. The identification of the true mountain species should present little difficulty; but it is only fair to add that in certain areas of Scotland, especially Breadalbane and Clova, hybrids are very frequent.

1	Prostrate shrubs with creeping rhizome or creeping and rooting stem	2
	Small shrubs without creeping rhizome or rooting stem	
2	Leaves oval to lanceolate, rarely rounded or retuse at apex, with dense appressed silky hairs at least beneath	*S. repens*
	Leaves always orbicular or broadly oval, rounded or retuse at apex, glabrous or nearly so	3
3	Leaves bright green and shining on both sides, with veins rather prominent but not impressed above	*S. herbacea*
	Leaves dark green above, with veins very strongly impressed, glaucous beneath	*S. reticulata*
4	Leaves glabrous or nearly so, shining above	5
	Leaves, at least beneath, densely silky-tomentose	6
5	Leaves bright green and shining on both sides	*S. myrsinites*
	Leaves pale or glaucous beneath	*S. arbuscula*

6 Leaves chiefly elliptic to oblong, 2 - 4 times as long as broad, to 5 cm. in length *S. lapponum*

Leaves all ovate or obovate, 1½ - 2 times as long as broad, to 6.5 cm. in length *S. lanata*

Distribution

S. arbuscula: locally abundant on basic cliff-ledges, but restricted to Argyll and Perth.

S. herbacea: widely distributed and locally common on the tops of almost all the higher mountains of Scotland, Northern England, Wales and Ireland; sometimes also on rock-ledges. Descends to 300 feet in Sutherland.

S. lanata: very local and rare on damp basic rocks in the mountains of Perth, Angus and Aberdeen.

S. lapponum: on damp rocks, screes and stream banks in mountainous districts, very rare in the English Lake District, widely distributed but local on the mainland of Scotland from Stirling northwards.

S. myrsinites: local and rare on rocks in Dumfries and from Argyll, Perth and Angus northwards to Orkney. Descends to 300 feet in Sutherland (see Appendix III, fig. 7b).

S. reticulata: locally frequent on basic rocks in Argyll, Perth, Angus, Aberdeen and Inverness.

APPENDIX III: DISTRIBUTION MAPS

THE BRITISH DISTRIBUTION of twenty different mountain plants is shown on the sixteen maps in this Appendix. The maps have been prepared, largely from the existing data in the botanical literature and in some of the principal Herbaria, in conjunction with the Distribution Maps Scheme of the Botanical Society of the British Isles. Each dot on the map signifies at least one reliable record for the plant concerned in a particular 10 kilometre square of the Ordnance Survey National Grid, which is marked round the margin of the map. The 10 kilometre Grid Reference for the particular record or records can thus be read back from the map, if required. No attempt has been made to distinguish on the maps between old and recent records, except in special cases where it is known that the plant is extinct in a particular locality. Most herbarium and literature records for the species mapped are nineteenth century ones, but there is good reason to think that such records for mountain plants are still valid today, for the mountain vegetation has been less drastically modified by man than the lowland communities.

The maps of *Hieracium* (J.R.) and *Alchemilla* (M.W.) are based solely on specimens personally checked by the author concerned. The existing literature records for such critical species are so confused as to be useless for this purpose, and it proved better to start again.

The probable completeness of the maps varies roughly in proportion to the rarity of the species, an apparent paradox which is easily understood when it is remembered that until the development of ecological studies, nearly all the emphasis in collecting and recording was placed on the rare or the unusual. Relatively widespread and locally frequent mountain plants such as the holly fern (*Polystichum lonchitis*, map 7a) are therefore somewhat under-recorded, whilst for the rare and local species, such as the two sandworts *Minuartia rubella* and *M. stricta*, the maps show the total known British distribution, and a new locality would be a first-class botanical discovery.

The Irish records on the maps have been contributed by Professor

D. A. Webb of Trinity College, Dublin, who is responsible for Irish data for the B.S.B.I. Maps Scheme. The slight error involved in placing these records—which are based upon a special extension of the Ordnance Survey National Grid over Ireland—or to the 'New Naturalist' base-map proved to be insignificant.

The authors would welcome reliable localised records which might appear to provide the basis for 'new dots' on the maps, and particularly such records for any vice-counties mentioned in the legend under the appropriate map but omitted from the map itself because of inadequate information as to locality, or doubt as to identification.

ADDENDUM

The 'Atlas of the British Flora' containing maps of the distribution of all British vascular plants, was published in 1962 by the Botanical Society of the British Isles, with Thomas Nelson and Sons Ltd. It is interesting to see that significant differences between the maps of mountain plants published in this Appendix and the Atlas are in fact only to be found in the relatively widespread species, holly fern (*Polystichum lonchitis*) and alpine saxifrage (*Saxifraga nivalis*), for both of which a number of new records have been collected in the Maps Scheme between 1955 and 1960.

M. W.
1965

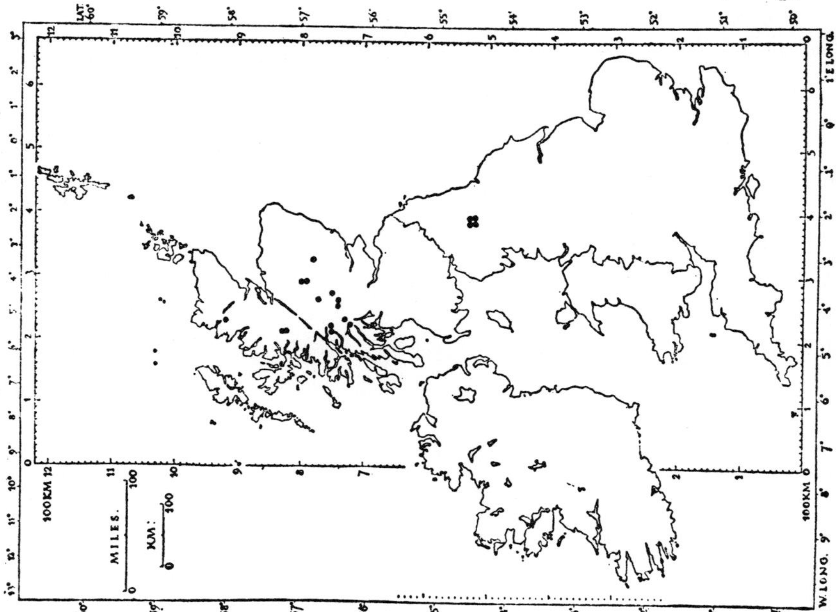

1a *Alchemilla glomerulans*

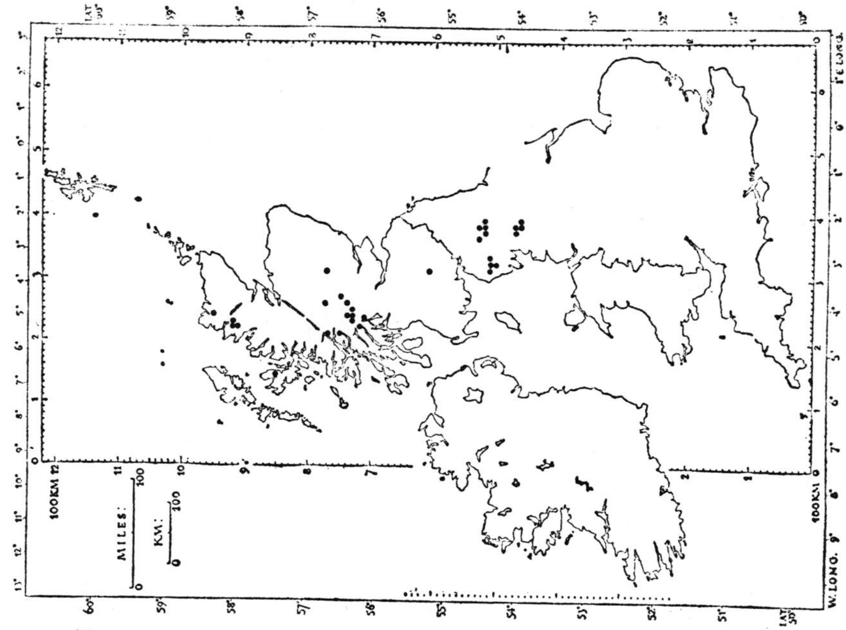

b *Alchemilla wichurae*

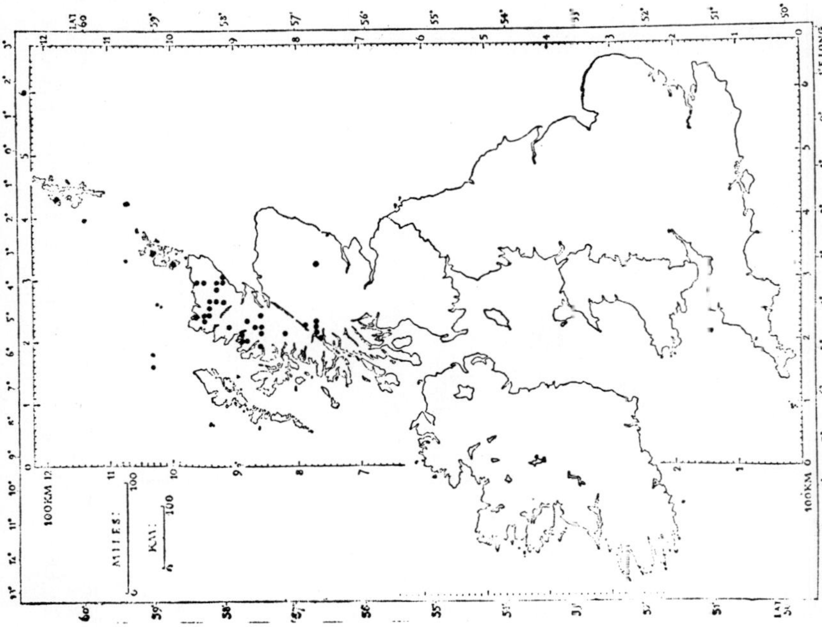

2a *Arctous alpina.* (Recorded also for Mid Perth (v.c. 88) and S. Aberdeen (v.c. 92); not recently confirmed for Angus (v.c. 90))

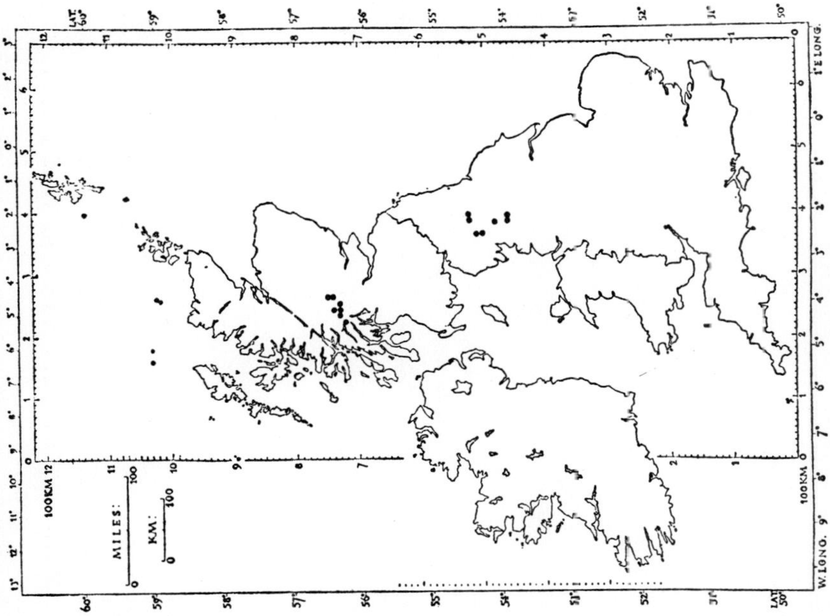

b *Bartsia alpina*

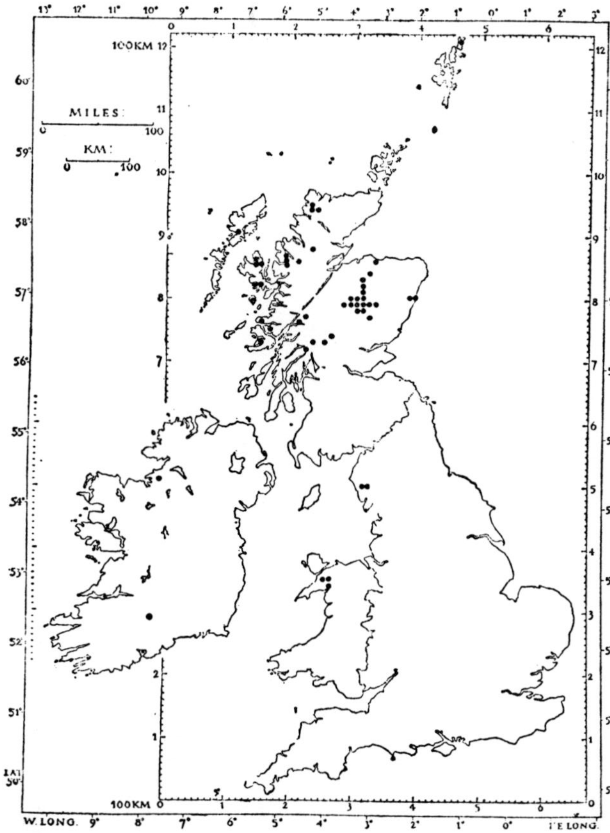

3a *Cardaminopsis petraea.* (Recorded also for E.Sutherland (vc. 107))

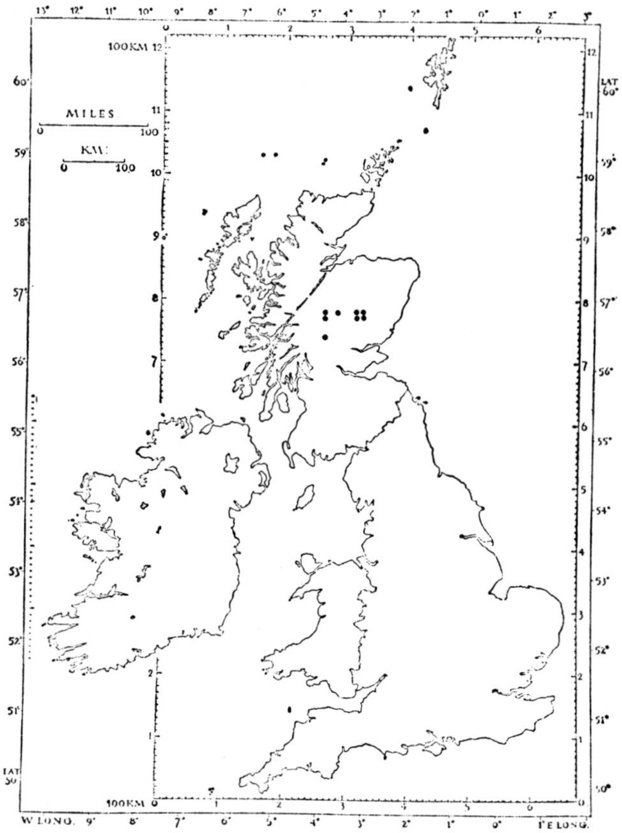

b *Carex rariflora.* (Recorded also for Banff (v.c. 94))

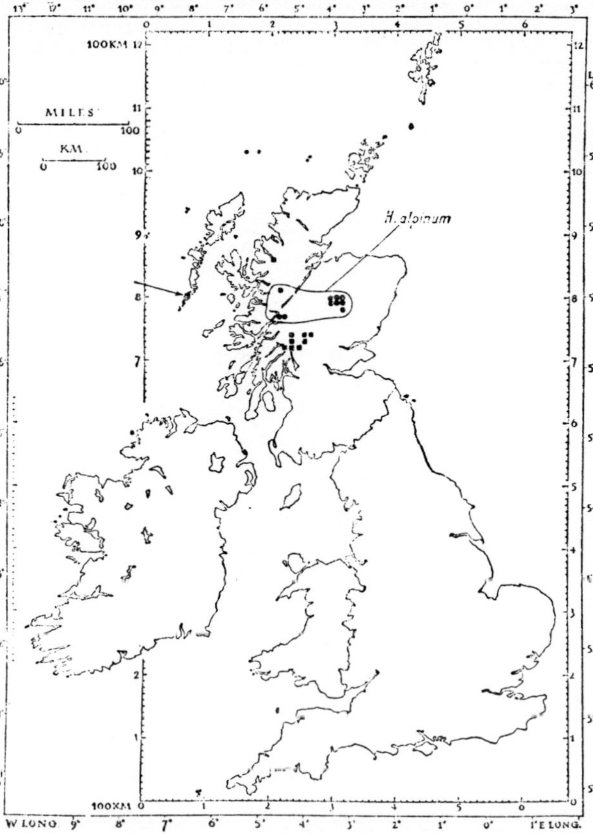

4a *Hieracium alpinum* (•) and *H. sinuans* (■)

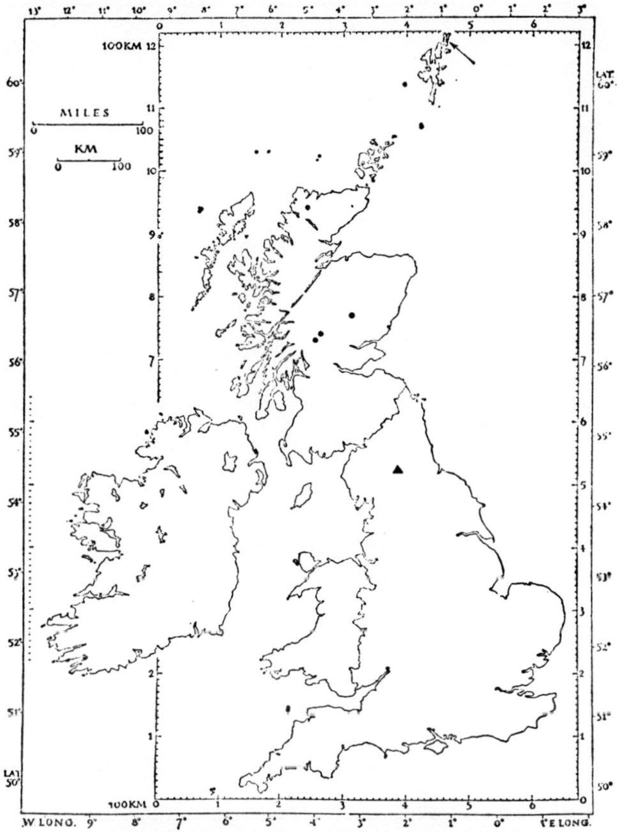

b *Minuartia rubella* (•) and *M. stricta* (▲) (*M. rubella* recorded also for E. Sutherland (v.c. 107))

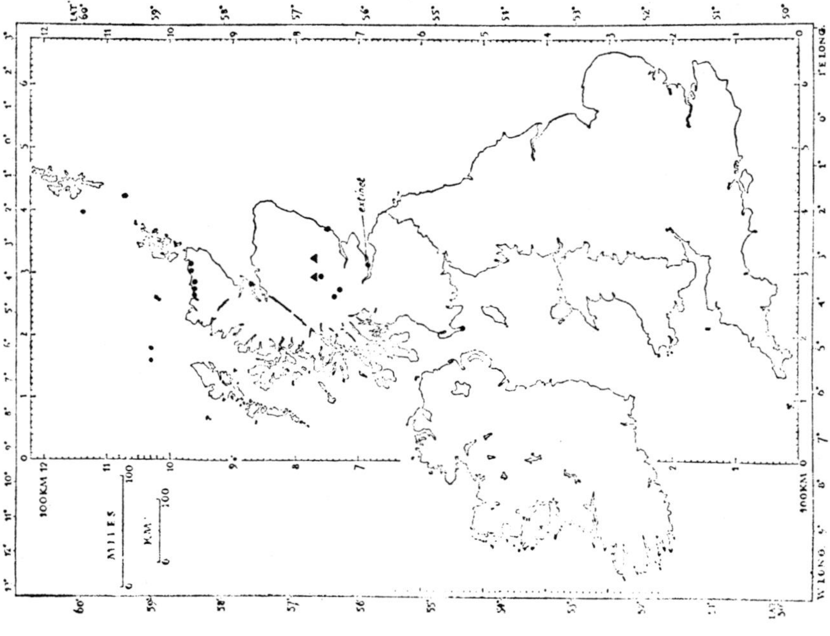

5*a* *Oxytropis halleri* (•) and *O. campestris* (▲). (*O. halleri* recorded also for Argyll (v.c. 98))

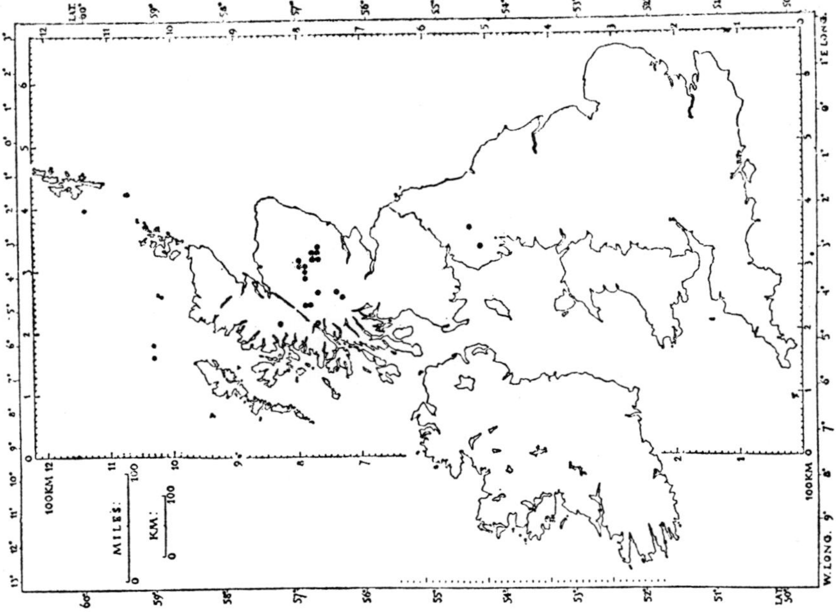

b. Phleum commutatum

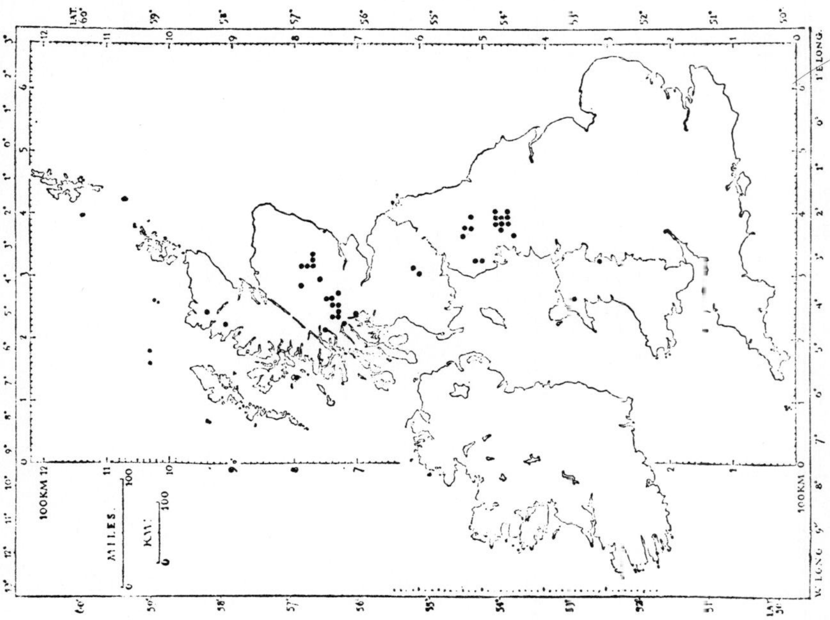

6a *Potentilla crantzii.* (Recorded also for Banff (v.c. 94), Westerness (v.c. 97) and E. Sutherland (v.c. 107))

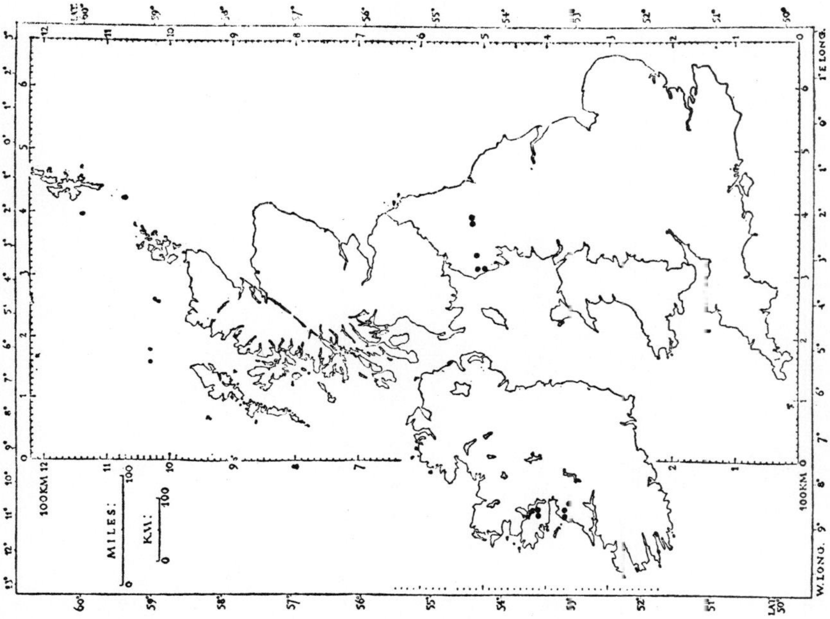

b *Potentilla fruticosa*

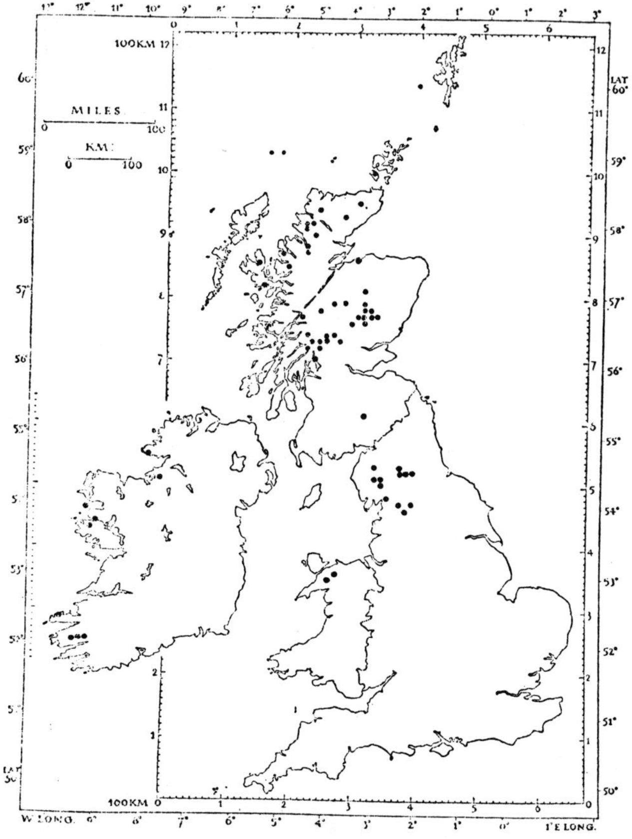

7a Polystichum lonchitis. (Recorded also for Merioneth (v.c. 48), Northumberland (v.c. 68) and Dumbarton (v.c. 99))

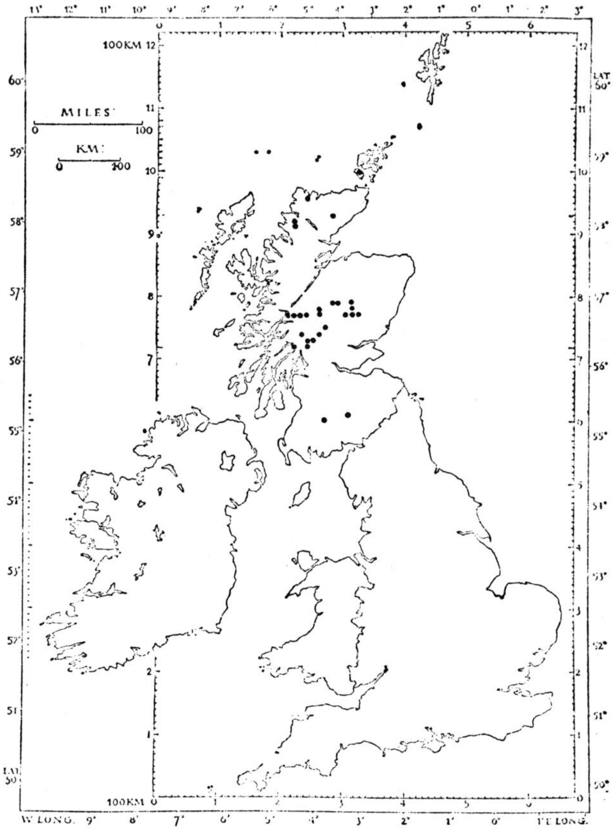

b Salix myrsinites. (Recorded also for Dumbarton (v.c. 99) and E. Ross (v.c. 106))

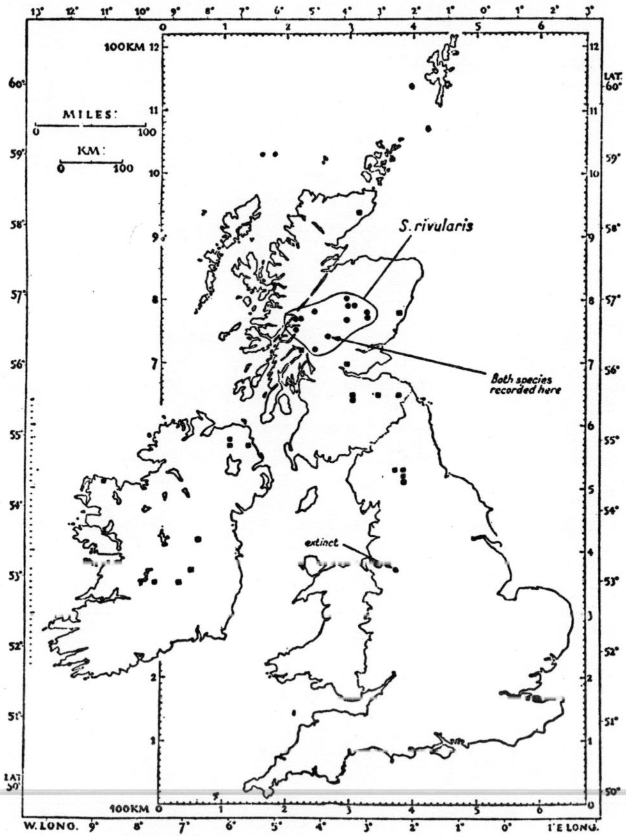

8a *Saxifraga rivularis* (●) and *S. hirculus* (■)

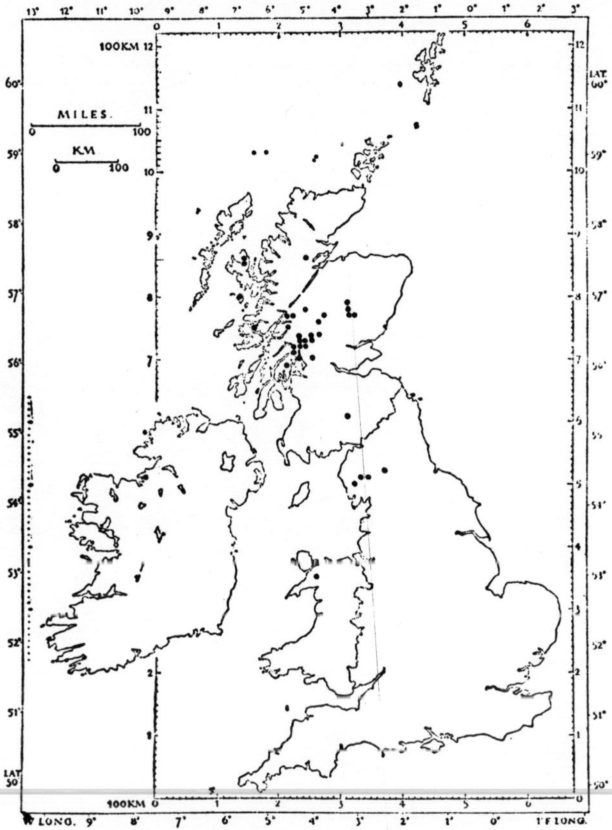

b *Saxifraga nivalis.* (Recorded also for Midlothian (v.c. 83) and Easterness (v.c. 96))

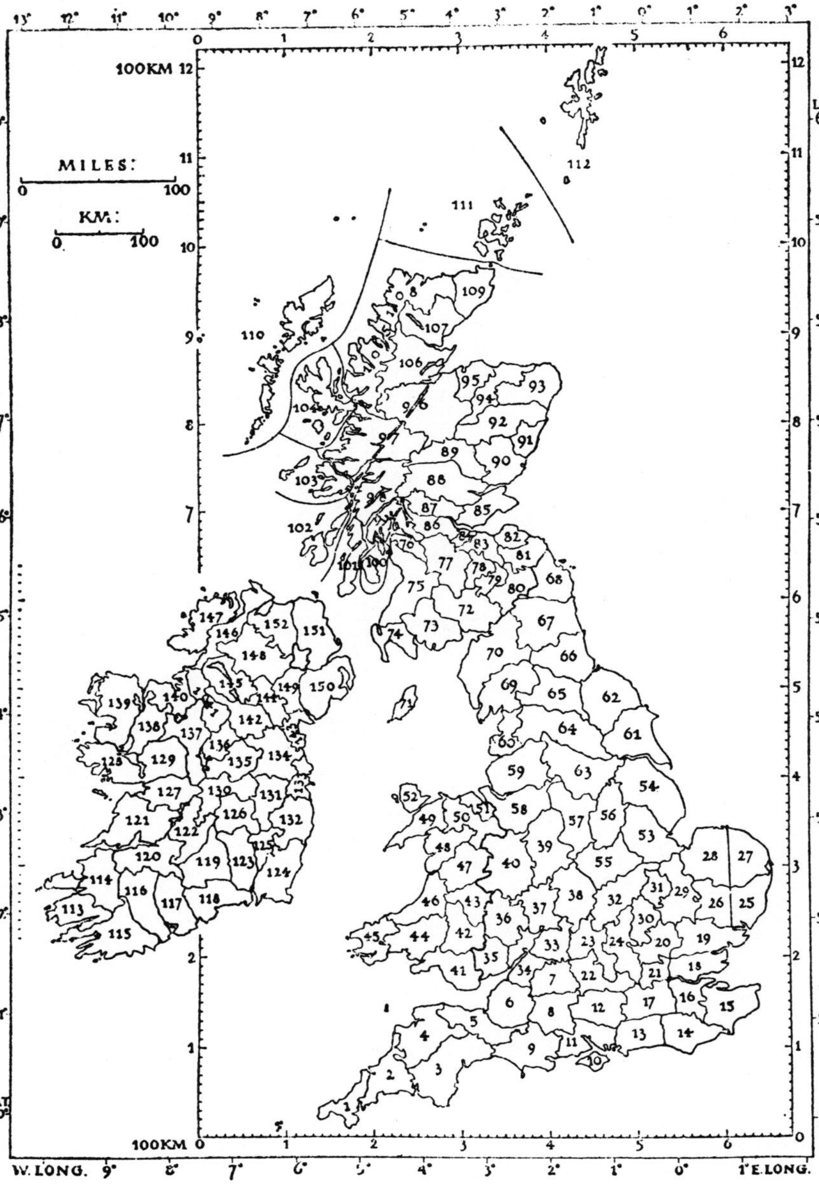
MILES:
0
100
KM:
0
100
100KM 12
100KM 0
LAT. 60°
LAT. 50°
W. LONG.
1° E. LONG.

REFERENCE TO VICE-COUNTIES

ENGLAND AND WALES

PENINSULA

1 West Cornwall with Scilly
2 East Cornwall
3 South Devon
4 North Devon
5 South Somerset
6 North Somerset

CHANNEL

7 North Wilts.
8 South Wilts.
9 Dorset
10 Isle of Wight
11 Hants. South
12 Hants. North
13 West Sussex
14 East Sussex

THAMES

15 East Kent
16 West Kent
17 Surrey
18 South Essex
19 North Essex
20 Herts.
21 Middlesex
22 Berks.
23 Oxford
24 Bucks.

ANGLIA

25 East Suffolk
26 West Suffolk
27 East Norfolk
28 West Norfolk
29 Cambridge
30 Bedford and detached part of Hunts.
31 Hunts.
32 Northampton

SEVERN

33 East Gloucester
34 West Gloucester
35 Monmouth
36 Hereford
37 Worcester
38 Warwick
39 Stafford and Dudley
40 Shropshire

SOUTH WALES

41 Glamorgan
42 Brecon
43 Radnor
44 Carmarthen
45 Pembroke
46 Cardigan

NORTH WALES

47 Montgomery
48 Merioneth
49 Carnarvon
50 Denbigh and parts of Flint
51 Flint
52 Anglesey

TRENT

53 South Lincoln
54 North Lincoln
55 Leicester with Rutland
56 Nottingham
57 Derby

MERSEY

58 Cheshire
59 South Lancashire
60 Mid Lancashire

HUMBER

61 South-east York
62 North-east York
63 South-west York
64 Mid-west York
65 North-west York

TYNE

66 Durham
67 Northumberland, South
68 Cheviotland, or Northumberland, North

LAKES

69 Westmorland with North Lancashire
70 Cumberland
71 Isle of Man

SCOTLAND

W. LOWLANDS

72 Dumfries
73 Kirkcudbright
74 Wigtown
75 Ayr
76 Renfrew
77 Lanark and E. Dumbarton

E. LOWLANDS

78 Peebles
79 Selkirk
80 Roxburgh
81 Berwick
82 East Lothian
83 Midlothian
84 West Lothian

E. HIGHLANDS

85 Fife with Kinross
86 Stirling
87 South Perth with Clackmannan, and parts of Stirling
88 Mid Perth
89 North Perth
90 Angus or Forfar
91 Kincardine
92 South Aberdeen
93 North Aberdeen
94 Banff
95 Moray or Elgin
96 Easterness (East Inverness with Nairn)

W. HIGHLANDS

97 Westerness (West Inverness with North Argyll)
98 Argyll (Main)
99 Dumbarton (West)
100 Clyde Isles
101 Cantire
102 South Ebudes (Islay, etc.) and Scarba
103 Mid Ebudes (Mull, etc.)
104 North Ebudes (Skye, etc)

N. HIGHLANDS

105 West Ross
106 East Ross
107 East Sutherland
108 West Sutherland
109 Caithness

NORTH ISLES

110 Outer Hebrides
111 Orkney
112 Shetland

IRELAND

113 South Kerry
114 North Kerry
115 West Cork
116 Mid Cork
117 East Cork
118 Waterford
119 South Tipperary
120 Limerick
121 Clare with Aran Isles
122 North Tipperary
123 Kilkenny
124 Wexford
125 Carlow
126 Leix
127 South-east Galway
128 West Galway
129 North-east Galway
130 Offaly
131 Kildare
132 Wicklow
133 Dublin
134 Meath
135 Westmeath
136 Longford
137 Roscommon
138 East Mayo
139 West Mayo
140 Sligo
141 Leitrim
142 Cavan
143 Louth
144 Monaghan
145 Fermanagh
146 East Donegal
147 West Donegal
148 Tyrone
149 Armagh
150 Down
151 Antrim
152 Derry

BIBLIOGRAPHY

ARNOTT, G. A. W. *See* HOOKER, W. J.

BACKHOUSE, J. 1856. Monograph of the British *Hieracia*. York.

BAILEY, K. 1833. The Irish Flora. Dublin.

BALFOUR, J. H. 1848. Notes of a Botanical Excursion, with Pupils, to the Mountains of Braemar, Glenisla, and Clova, and to Ben Lawers. *Edin. Philos. J. 45*, 122.

BALME, O. E. 1954. *Viola lutea* Huds. in Biological Flora of British Isles. *J. Ecol. 42*, 234.

BARRINGTON, R. M. 1908. Henry Chichester Hart. *Irish Naturalist, 17*, 249.

BAUHIN, C. 1620. Prodromus Theatri botanici. Frankfurt.

BENTHAM, G. and HOOKER, J. D. 1924 (1947). Handbook of the British Flora. Ed. 7, rev. by A. B. Rendle. London.

BLAKELOCK, R. A. 1953. *Artemisia norvegica* Fries in Scotland. *Kew Bulletin, 1953*, 173.

BURGES, A. 1951. The Ecology of the Cairngorms. Part III. The *Empetrum-Vaccinium* Zone. *J. Ecol. 39*, 271.

BURTT, B. L. 1950. *Koenigia islandica* in Britain. *Kew Bulletin, 1950*, 266.

CLAPHAM, A. R., TUTIN, T. G. and WARBURG, E. F. 1952. Flora of the British Isles. Cambridge.

COLGAN, N. 1900. Botanical Notes on the Galway and Mayo Highlands. *Irish Naturalist, 9*, 111.

—— 1904. Flora of the County of Dublin. Dublin.

CRAMPTON, C. B. 1911. Vegetation of Caithness. Edinburgh.

DAHL, E. 1951. On the Relation between Summer Temperature and the Distribution of Alpine Vascular Plants. *Oikos 3*, 22.

DARWIN, C. 1859. Origin of Species. London.

—— ed. Darwin, F. 1909. Foundations of the Origin of Species ; Two Essays. Cambridge.

DARWIN, F. (ed.) 1888. Life and Letters of Charles Darwin. London.

DAVIES, E. 1953. Notes on Two Rare Scottish Sedges. *Watsonia, 2,* 300.

DE L'ECLUSE, C. 1583. Rariorum aliquot stirpium per Pannoniam, Austriam, et vicinas quasdam provincias observatarum historia . . . Antwerp.

DILLENIUS, J. J. 1724. *See* RAY, J.

DRUCE, G. C. 1912. George Don. *Brit. Assn. Handbook, Dundee Meeting, 1912.*

—— 1920. Dubious Plants of Britain. *B. E. C. Rep. 1919,* 751.

—— 1923. *Carex microglochin. B. E. C. Rep. 1923,* 68.

—— 1932. The Comital Flora of the British Isles. Arbroath.

DRUMMOND, J. L. 1829. Memoir of the late John Templeton Esq. *Loudon's Mag. of Nat. Hist. 1,* 403.

DU RIETZ, G. E. 1940. Problems of Bipolar Distribution. *Acta Phytogeographica Suecica, 13,* 215.

EVANS, E. PRICE. 1932. Cader Idris: a Study of certain Plant Communities in South-west Merionethshire. *J. Ecol. 20,* 1.

FORBES, E. 1846. On the Connexion between the Distribution of the existing Fauna and Flora of the British Isles, and the Geological Changes which have affected their area, especially during the epoch of the Northern Drift. *Mem. Geol. Survey U.K. 1,* 336.

FORD, E. B. 1955. Moths. London.

FOTHERGILL, P. G. 1941. Studies in *Viola, III. New Phyt. 40,* 139.

GEIKIE, A. 1876. Notes of a Botanico-Geological Trip to Clova. *Trans. Bot. Soc. Ed. 12,* 159.

GERARD, J. 1597. The Herball. London.

GILMOUR, J. S. L. and WALTERS, S. M. 1954. Wild Flowers. London.

GODWIN, H. 1949. The Spreading of the British Flora, Considered in Relation to Conditions of the Late-Glacial Period. *J. Ecol. 37,* 140.

—— 1954. *Koenigia islandica* in the Scottish Late-Glacial. *Proc. B.S.B.I. 1,* 86.

—— 1956. History of the British Flora. Cambridge.

GOOD, R. D'O. 1928. On the Distribution of the Lizard Orchid (*Himantoglossum hircinum* Koch). *New Phyt. 35,* 142-170.

GOODWAY, K. M. 1955. *Proc. B.S.B.I. 1,* 383

GORDON, H. SETON. 1950. Snow Flora of the Scottish Hills. *Nature, 165,* 132.

GREGOR, J. W. 1938. Experimental Taxonomy II. Initial Population Differentiation in *Plantago maritima* L. of Britain. *New Phyt. 37*, 15.

GUNTHER, R. T. 1945. Early Science in Oxford, Vol. XIV. Life and Letters of Edward Lhwyd. Oxford.

GUSTAFSSON, A. 1946. Apomixis in Higher Plants. Lund.

HANBURY, F. J. 1889-98. Illustrated Monograph of the British *Hieracia*. London.

HANCOCK, T. W. 1870. History of Llanrhaiadr-ym-Mochnant. Montgomeryshire Collections, Vols. 4 and 5.

HEGI, G. 1906-39. Illustrierte Flora von Mittel-Europa, 7 vols. Munich.

HESLOP-HARRISON, J. 1947. Recent Researches on Flora and Fauna of the W. Isles of Scotland and their Biogeographical Significance. *Proc. Belfast Nat. Hist. and Phil. Soc. Ser. 2, 3, 2*, 87.

—— 1953. The North American and Lusitanian Elements in the Flora of the British Isles; *in* The Changing Flora of Britain, ed. J. E. Lousley (p. 105). Arbroath.

HOLDEN, A. E. 1952. Plant Life in the Scottish Highlands. Edinburgh.

HOOKER, J. D. 1861. Outlines of the Distribution of Arctic Plants. *Trans. Linn. Soc. 23*, 251.

HOOKER, W. J. 1830. British Flora. London.

—— and ARNOTT, G. A. W. 1850. British Flora, Ed. 6. London.

HOW, W. 1650. Phytologia Britannica. London.

HUDSON, W. 1762. Flora Anglica. London.

HULTEN, E. 1937. Outline of the History of Arctic and Boreal Biota during the Quaternary Period. Stockholm.

——1954. *Artemisia norvegica* Fr. and its Allies. *Nytt Magasin for Botanikk, 3*, 63.

JOHNSON, T. 1634 and 1641. Mercurius Botanicus, Parts I and II. London.

LEACH, W. 1930. A Preliminary Account of the Vegetation of some non-calcareous British Screes. *J. Ecol. 18*, 321.

LEES, F. A. 1888. The Flora of West Yorkshire. London.

LHWYD, E. *See* GUNTHER, R. T.

LIGHTFOOT, J. 1777. Flora Scotica. London.

LINTON, W. R. 1905. Account of the British *Hieracia*. London.

LOUSLEY, J. E. 1950. Wild Flowers of Chalk and Limestone. London.

LØVE, A. 1953. Subarctic Polyploidy. *Hereditas, 39*, 113-124.

MACGILLIVRAY, W. 1832. Remarks on the Phanerogamic Vegetation of the River Dee. *Memoirs of the Wernerian Nat. Hist. Soc. 6.*

MACKAY, J. T. 1825. Catalogue of the Plants found in Ireland. Dublin.

—— 1836. Flora Hibernica. Dublin.

MANLEY, G. 1952. Climate and the British Scene. London.

MANTON, I. 1950. Problems of Cytology and Evolution in the Pteridophyta. Cambridge.

MARSHALL, E. S. 1888. Notes on Highland Plants. *J. Bot. 26,* 149.

—— 1894. On an apparently undescribed *Cochlearia* from Scotland. *J. Bot. 32,* 289.

MELDERIS, A. 1953. Some Parallels between the British and Scandinavian Mountain Floras *in* The Changing Flora of Britain, ed. J. E. Lousley (p. 89). Arbroath.

MERRET, C. 1666. Pinax Rerum Naturalium Britannicarum. Ed. 2, 1667. London.

METCALFE, G. 1950. The Ecology of the Cairngorms. Part II, The Mountain *Callunetum. J. Ecol. 38,* 46.

NEILL, P. 1809. An Account of a Forfar Garden. *Scots Magazine* for June 1809.

NORDHAGEN, R. 1932. Studien über die Skandinavischen Rassen des *Papaver radicatum* Rottb . . . *Bergens Mus. Arbok 1931, Naturvid. rekke 2.*

PEARSALL, W. H. 1950. Mountains and Moorlands. London.

PENNANT, T. *See* LIGHTFOOT, J.

PIGOTT, C. D. 1954. Species Delimitation and Racial Divergence in British *Thymus. New Phyt. 53,* 470.

—— 1955. *Thymus drucei* in Biological Flora of British Isles. *J. Ecol. 43,* 377.

PRAEGER, R. Ll. 1934. The Botanist in Ireland. Dublin.

—— 1950. The Natural History of Ireland. London.

PUGSLEY, H. W. 1948. A Prodromus of the British *Hieracia. J. Linn. Soc. (Bot.), 54,* 1.

RAY, J. 1670. Catalogus Plantarum Angliae. Ed. 2, 1677. London.

—— 1688. Fasciculus Stirpium Britannicarum post editum Plantarum Angliae Catalogum observatarum. London.

—— 1690. Synopsis methodica Stirpium Britannicarum. Ed. 2, 1696. Ed. 3 (by Dillenius), 1724. London.

Reid, E. M. 1949. The Late-Glacial Flora of the Lea Valley. *New Phyt. 48*, 245.

Salisbury, E. J. 1942. The Reproductive Capacity of Plants. London.

—— and Wilmott, A. J. 1935. *In* Discussion on the Origin and Relationship of the British Flora. *Proc. Roy. Soc. Ser. B. 118,* 197.

Sandwith, N. Y. 1952. A Mysterious *Carex* on Ben Lawers. *Watsonia, 2*, 279.

Schroeter, C. 1926. Das Pflanzenleben der Alpen. Zurich.

Sibbald, R. 1684. Scotia illustrata. Edinburgh.

Smith, C. 1746. The antient and present state of the County of Waterford. Dublin. Ed. 2, 1774.

—— 1750. The antient and present state of the County and City of Cork. Dublin. Ed. 2, 1774.

—— 1756. The antient and present state of the County of Kerry. Dublin. Ed. 2, 1774.

Smith, J. E. 1790-1814. English Botany. Ill. J. Sowerby. 36 vols. London.

Smith, R. 1900. Botanical Survey of Scotland, Part I, Edinburgh District; Part II, North Perthshire. *Scot. Geog. Mag.* 1900.

Smith, W. G. 1904-5. Botanical Survey of Scotland, Parts III and IV, Forfar and Fife. *Scot. Geog. Mag. 20 and 21.*

Syme, J. T. I. Boswell 1863-92. English Botany. Ed. 3. London.

Tansley, A. G. (ed.) 1911. Types of British Vegetation. Cambridge.

Tansley, A. G. 1939. The British Islands and their Vegetation. Cambridge.

Threlkeld, C. 1727. Synopsis Stirpium Hibernicarum. Dublin.

Turrill, W. B. 1938. Material for a study of taxonomic problems in *Taraxacum. B.E.C. Rep.* 1937, 570.

—— 1948. British Plant Life. London.

Wade, W. 1794. Catalogus systematicus plantarum indigenarum in comitatu Dublinensi inventarum. Part 1. Dublin.

—— 1804. Plantae Rariores in Hibernia Inventae. Dublin.

Walker, S. 1955. Cytogenetic Studies in the *Dryopteris spinulosa* Complex. *Watsonia 3,* 193.

Walters, S. M. & Lambert, C. 1956. A Reconsideration of *Silene caelata* C. Reid. *New Phyt. 54.*

Watson, H. C. 1832. Outlines of the Geographical Distribution of British Plants. Edinburgh.

—— 1847-59. Cybele Britannica. London.

—— 1873-4. Topographical Botany. Thames Ditton.

—— 1883. Topographical Botany. Ed. 2. London.

WATT, A. S. and JONES, E. W. 1948. The Ecology of the Cairngorms. Part I. The Environment and the Altitudinal Zonation of the Vegetation. *J. Ecol. 36*, 283.

WEBB, D. A. 1951. The Mossy Saxifrages of the British Isles. *Watsonia, 2*, 22.

WHITE, F. BUCHANAN. 1879. The Mountain Lepidoptera of Britain. *Scot. Nat. 5*, 97 and 149.

—— 1885. *Myosotis alpestris* in Forfarshire. *J. Bot. 23*, 26.

—— 1898. Flora of Perthshire. Edinburgh.

WILLIS, J. C. and BURKILL, I. H. 1895-1908. Flowers and Insects in Great Britain. Part I, 1895. *Ann. Bot. 9*, 227; Part II, 1903. *Ann. Bot. 17*, 313; Part III, 1903. *Ann. Bot. 17*, 539; Part IV, 1908. *Ann. Bot. 22*, 603.

WINCH, N. 1819. An Essay on the Geographical Distribution of Plants through the Counties of Northumberland, Cumberland and Durham.

WYCHERLEY, P. R. 1953a. The Distribution of the viviparous Grasses in Britain. *J. Ecol. 41*, 275.

—— 1953b. Proliferation of Spikelets in British Grasses. *Watsonia 3*, 41.

YEO, P. F. 1956. Hybridisation between diploid and tetraploid species of *Euphrasia*. *Watsonia* 4.

INDEX

All page-references to plants are given under their Latin names. English names are indexed, with few exceptions, under the name of the group, e.g. 'alpine saxifrage' under 'saxifrage', and are followed by their Latin equivalents, under which page-references will be found. Books are indexed under authors, not titles. Place-names have been indexed if sufficiently precise, but counties, for example, have not been included.

Important page references are in italics; numbers in bold type refer to pages opposite to which photographic illustrations will be found.